CAGE & AVIARY SE[

Mynah Birds
&
Related Starlings

MYNAH BIRDS & RELATED STARLINGS

JAMES BLAKE

Beech Publishing House
7, Station Yard
Elsted Marsh
MIdhurst   GU29 0JT
West Sussex, England

© James Blake, 2002

First Edition 2002

ISBN 1-85736-291-8

This book is copyright and may not be reproduced or stored in any way without the express permission of the publishers in writing

***A Catalogue entry for
this book is at
the British Library***

**Beech Publishing House
7, Station Yard
Elsted Marsh
MIdhurst   GU29 0JT
West Sussex, England**

FOREWORD

Interest in aviary birds continues unabated and therefore it is a pleasure to add one more to our list. These birds give many hours of pleasure to those who keep them.

Mynah birds are boisterous, interesting companions, and are part of the large Starling Family. They breed in captivity, provided a large flight is provided and they are not in a mixed collection of birds, although it must be added that they have been known to breed when they are housed with other species. Usually, though, if serious breeding is to be considered there must be the correct conditions. Moreover, with the presence of other birds both eggs and chicks may be destroyed.

This is a basic handbook, intended for those who wish to know the fundamentals of bird keeping, relating to Starlings and Mynah birds.

JB   March, 2002

CONTENTS

1. Background	1
2. Natural History	7
3. Aviaries & Cages	47
4. Feeding	63
5. Case Studies	75

COLOURED PLATE

The plate overleaf indicates the variety of Starlings which may be found. The Mynah Bird is a popular variety.

Starlings & Mynahs
From *Birds of Burma* (B E Smythies)
1. Common Mynah
2. Hill Mynah (Gracula religiosa intermedia)
3. Black Collared Starling
4. Vinous-Breasted Starling
5. Chestnut Tailed Starling
6. Asian Pied Starling
7. Jungle Mynah

1
BACKGROUND

• •

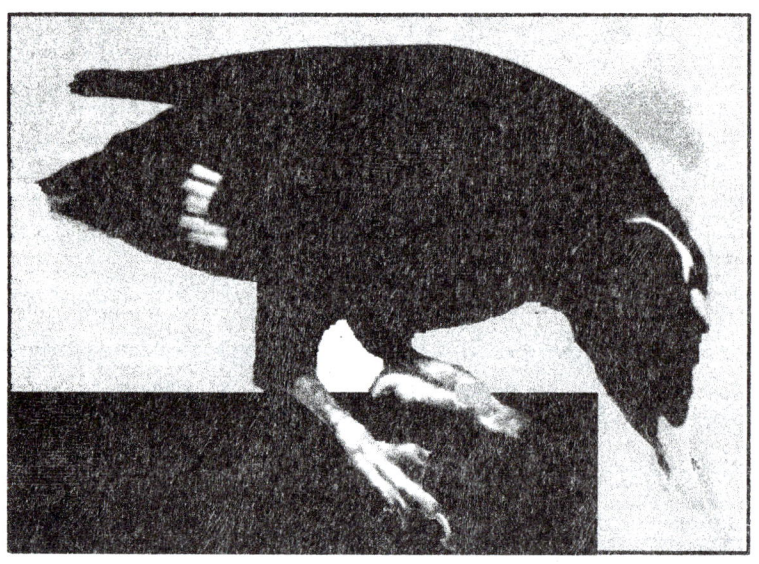

A Typical Pose

THE FAMILY

The family into which the Mynahs belong is the Starling Group which are part of the ORDER **Passeriformes** and the FAMILY *Sturnidae*. They come from Africa, India and South-east Asia. Unfortunately, they are quite a mixture, which does not lead to easy understanding when viewed as a whole.

Mynahs, the subject of this book, are also under different species names, although for most practical purposes the Greater Hill Mynah is the most popular and is widely kept. This goes under the name of Gracula religiosa. It comes from Indonesia and South-east Asia. It is about 10 inches (25cm) in length and is a strong, active bird.

Main Species

The main species are:

1. Mandarin Mynah *(Sturnus sinensis)* This is also known as the Chinese Mynah. Distribution: China, Formosa, and South-east Asia.

2. Pied Mynah *(Sturnus c. contra)* This comes from Assam, Northern India and Nepal. It measures around 9 inches (23 cm) and is well balanced in appearance with a long beak.

3. Andaman Mynah *(Sturnus erythropygius adamamensis)* Also known as Andaman Starling or White-headed Starling. Comes from the Andaman and Nicobar Islands. Size around 8 inches (20 cm). They have been bred in captivity.

4. Pagoda Mynah *(Sturnus pagodarum)*. Found in Afghanistan, Nepal, India and Sri Lanka. Size 8 inches (20 cm). Also called the **Pagoda Starling** and **Brahminy Mynah**. They have been bred in captivity and are relatively quiet in behaviour. The breast is a bay/buff colour and is most attractive.

5. Rothchild's Grackle *(Leucospar rothchildi)*. This is from the Island of Bali in Indonesia. It is a large, well developed bird with a crest, which makes it one of the most outstanding of the Mynahs. They have been bred in captivity, but they are difficult.

6. Common Mynah *(Acridotheres t. tristis)*. Originally from India and Afghanistan they have now settled in New Zealand, Australia and Malaysia. Again around 9 inches in length. They can be bred in aviaries and many successes have been recorded.

7. Brown Mynah *(Acridotheres f. fuscus)*. From Burma, India, Malaysia, Java and the Celebes Islands. Only few successes in breeding. Similar in size to the Common Mynah.

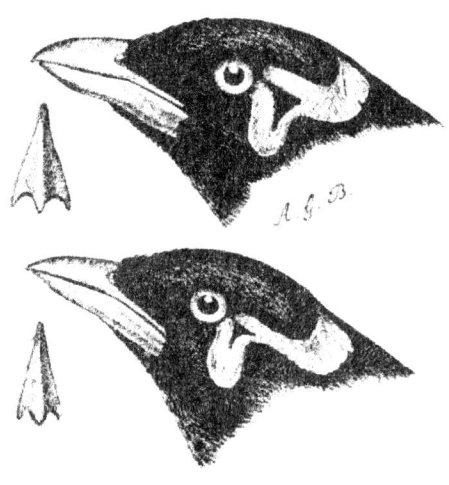

Male and Female of the Hill Mynah (possibly *intermedia*)
Drawn by Dr A G Butler a pioneer bird keeper

8. Bank Mynah *(Acridotheres ginginianus)*. From India and similar to the preceding two species in habits, but slightly smaller.

9. Chinese Crested Mynah *(Acridotheres cristatellus)*. Their natural habitat is in South China, Burma, Formosa, Laos, Vietnam, and Hainan. They are quite large, being about 10.50 inches (26.7 cm) in length. They have been domesticated for a considerable period being kept as early as 1840 in Amsterdam.

10. Golden Crested Grackle (Ampeliceps *coronatus)*. From Burma, India, Laos, Thailand, Khmer, Vietnam and Malaysia. It is smaller than the previous species being about 8 inches long (20.3 cm). This can be recommended as an aviary bird because they become quite tame.

11. Coleto Mynah *(Sarcops c. calvus)*. Also known as the Bald-headed Starling. They come from the Philippines. In length they are about 11 inches (28 cm) and can be bred from if kept in a separate aviary for the breeding pair. They can be taught to speak.

12. Ceylon Mynah *(Gracula ptilogenys)*. This comes from Sri Lanka. The male is 9.50 inches (24 cm) and the female slightly smaller. They will do well in aviaries. They do not possess wattles or lobes like the others.

13. Lesser Indian Hill Mynah *(Gracula religiosa indica)*. Also known as the Southern Hill Mynah, it comes from Southern India and Sri Lanka. It is about 9.50 inches (24 cm). They have been bred in captivity, but are not as able as the other species in this category.

14. Indian Mynah *(Gracula religiosa intermedia)* . From India, Burma, southern China and the hills of the southern Himalayas. Size: around 11 inches (28 cm).

15. Greater Hill Mynah (*(Gracula r. religiosa)*. From Java, Bali, Malaysia and Sumatra and adjacent islands. Size 12 inches (30cm).

This is the most popular of the Mynahs because it imitates the human voice extremely well. They are to be found in many zoological gardens and bird gardens as well as being kept in cages as a talking bird.

Green Glossy Starlings
These are related to the Mynah Birds

2

NATURAL HISTORY

Pair Of Greater Hill Mynahs

EARLY HISTORY*

The Mynah bird (originally spelt *Mina* or *Mino*) has a long history, being recorded from the early days of natrural history writing. In this chapter, an extension of Chapter 1, some of the more important recordings are summarized:

On average the Mynah bird** is about ten inches long, three and a half of which belong to the tail; the wing measures five inches and a quarter. The feathers upon the head, nape, and breast are of a brilliant black; the rest of the coat is reddish brown, the wings and back being of a deeper shade, and the underside lighter than the rest of the body; the exterior quills are black, but white at the root, thus giving a somewhat spotted appearance to the wing; the tail is black, and tipped with patches of white, the latter becoming gradually wider towards the sides; the belly and lower wing-covers are also white.

The Mynahs are among the commonest birds in India, Assam, and Burmah, where they frequent the neighbourhood of towns and villages in preference to more wooded districts. A tree is usually selected as their sleeping-place, and from this point they fly over the country in small parties in search of food, stealing occasionally even into the huts of the natives, in order to obtain cooked rice, of which they are very fond; some follow the flocks and herds, and seize the grasshoppers as they rise from the grass when disturbed by the cattle, others seek subsistence by plundering the gardens and orchards in their vicinity. When upon the ground the Minah walks with ease, constantly bowing its head as it goes, and occasionally springing to a considerable distance; its flight is heayy, direct, and tolerably rapid, and its notes rich and varied. So little fear is exhibited by these birds that they build almost exclusively in the vicinity of houses, or even in temporary cages that are hung out for their accommodation. In Mosuri, where this species is only a summer visitor, it usually prefers making its nest within a hollow tree.

---

* Based on a history included in *Cassell's Book of Birds*, by various authorities.

** **Description for Common Mynah** *(Acridotheres t. tristis); it will be appreciated that a number do not fit into this general descrip[tion and some are quite different, eg, Rothschild's Grackle.*

Like the Starling, it easily acquires the art of speaking, and of imitating a variety of sounds. The Minah has been dedicated by the Indians to their god RAM, and is usually represented as perched upon his hand.

Major Norgate has given a full description of this interesting bird, from which our space will only allow us to extract the following account of its quarrelsome propensities and regular pitched battles, he tells us, are of constant occurrence amongst these pugnacious little creatures; the two combatants, who usually belong to different flocks, coming to the ground, in order the better to carry on their struggle, which is maintained by clawing, beating with the wings, and rolling round each other, screaming loudly as the combat waxes hot; only for a very brief space, however, is the fight confined to these two champions of the rival parties ; one after another the rest come down and mingle in the fray, which often rages so fiercely that broken wings or other injuries at last. compel the untiring combatants to cease their strife. The same writer describes the Minah's manner of singing as being very amusing: it inflates its chest as though about to make a most tremendous effort, and then gives voice to such a variety of crowing, grunting, and squeaking sounds as cannot fail to astonish its hearer. When in flight the notes of these birds are by no means unpleasing; but if alarmed their cry rises to a loud, hoarse shriek, the rest of the party usually joining chorus until the uproar becomes general. The nest is constructed with the utmost carelessness, and is, in fact, a mere heap of straw, twigs, rags, or even shreds of paper; but in spite of the discomfort of the home thus provided for the young, the latter are tended by both parents with great affection.

THE GRACKLES

These constitute a race of Starlings that have always been regarded with great favour by mankind. These birds are of a moderate size, with thick bodies, and short wings and tails; the beak, which equals the head in length, is thick, high, and in its transverse section of a square form, the upper mandible is rounded and much vaulted at its roof: The fourth quill of the wings is longer than the rest, and the tail, which is

Natural Habitat

rounded at its tip, is composed of twelve feathers; the feet are strong, and the head is furnished on each side with two moveable appendages resembling flaps of skin (which are usually brightly coloured) hanging down from behind the eyes. The plumage is soft, and of a satin-like brilliancy.

THE MUSICAL GRACKLE.

Gracula religiosa* is about ten inches long and eighteen and a half inches in breadth ; the tail measures nearly three inches, and the wing five inches and three-fifths.

The plumage of this species is of a uniform rich, deep, purplish black, shaded with green upon the lower part of the back and upper wing-coverts ; upon the under surface of the body this beautiful green shimmer is less distinctly visible; the wings and tail are jet black, the former edged with a white band, formed by a series of patches, with which the first seven primary quills are marked; the strange fleshy flaps to which we have alluded are of a bright yellow colour, and are appended behind the eyes, passing over the ears, at which part they become considerably dilated. A naked space under the eyes is also of a brilliant yellow. The beak is orange, the feet yellow, and the eyes dark brown.

Jerdon tells us that these birds principally inhabit the woods of Eastern India, and that they are found in considerable numbers in the Rhat Mountains and other elevated regions, living at an altitude of 3,000 feet above the level of the sea, and only making their appearance in large flocks during the winter; at other seasons of the year they are usually met with in parties of six or seven. These groups pass the night together, generally in beds of reeds or bamboo thickets .upon the banks of the mountain streamlets. Their food consists of various kinds of fruit and berries, and their visits are therefore greatly dreaded by the proprietors of fields and gardens. The Grackle is lively and active, much resembling the Common Starling in disposition: its song is cheerful and varied, but contains many unpleasing notes; its powers of imitation

---

* This is possibly the Indian Mynah *Gracula religiosa intermedia*, which is about 11 inches in length (9 cm), but there are some variations from the general description of this species.

are so highly developed as to render it a most interesting companion when tamed; indeed, some of the admirers of this gifted bird declare it to be superior to the Parrot in the art of mimicry, and at the same time entirely without the disagreeable noisy habits that often render the latter intolerable. When caged, the Grackle not only becomes much attached to those who feed it, but soon familiarises itself with all the dogs and cats of the establishment, and will even fly fearlessly about the house in search of food.

Our own experience does not allow us to speak in quite such unqualified terms of praise as the writer from whom we quote: we have seen an instance in which one of these birds was so voracious as scarcely to allow itself time to utter a sound and so pugnacious and quarrelsome as to be an object of dread to all its feathered companions who suffered severely from its beak and claws.

PRESENT POSITION ON GRACKLES

The Grackles are still regarded as the principal speaking birds, and the knowledge on them has been greatly increased. Moreover, it will be appreciated that there are many sub-species. Thus the Hill Mynah has at least 10 sub-species. There now follows a description of the main species.

It should be noted that describing with accuracy can be difficult and, inevitably, some variations will be found. Often the different size and shape of the wattles is the main determining factor of these birds which are best at imitating and talking. An example is below:

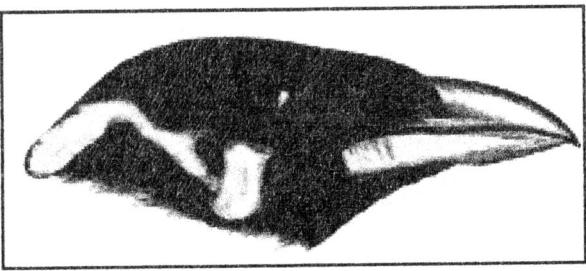

Head: *Intermedia*

1. Mandarin Mynah *(Sturnus sinensis)*

This species is from China, Formosa and South-east Asia. It has the tail and flight feathers black, but with white on tail and wings. The rest of the body is a greyish white, pale colour.

It has been described by Butler* as follows:

> Above ash-grey, somewhat more buffish on lower back; rump and upper tail-coverts; creamy-buff ; scapulars buffish-white; lesser and median wing-coverts white ; greater coverts creamy-buff ; remaining wing. feathers black, externally glossed with green or purple; inner secondaries wholly green; tail black glossed with green, the tips increasingly buff from centre to outer-most feather; crown creamy-buff. greyish on nape and back of neck; sides of head, chin, and throat buff, deeper in front, paler behind; sides and front of neck and breast pearl-grey, a few paler feathers on back of throat; abdomen greyish-white ; sides and flanks tawny, paler behind; thighs and under tail-coverts creamy-buff ; wing-coverts and axillaries pale salmon, white at base; flights below blackish, ashy on inner edge; bill blue tipped with yellow ; feet fleshy-grey; irides white (Oates), black (David), pearl-grey (Russ), who all says the bill is grey-green and the feet horn-yellow.
>
> Female not differentiated.

Apparentlly this species arrives in the South of China in the summer in multitudinous flocks, and always seeks the vicinity of human dwellings. It builds its nest in holes in the roofs. A very common spring visitor, staying to breed, and leaving about the end of September. The eggs are blue.

It was always a scarce bird and was rarely on the market. It has been bred in a number of zoos.

Foreign Birds for Cage & Aviary, A G Butler (nd)

Pied Mynah

2. Pied Mynah *(Sturnus c. contra)*

As described by Butler (ibid)

> Above, blackish-brown; scapulars white externally; -rump white; upper tail-coverts blackish-brown; lesser wing-coverts white; median and greater coverts with greenish margins; tips of primaries edged with white; secondaries more bronze-brown; tail black, fringed at tip with white; crown, nape, and back of neck greenish, black; feathers of forehead and eyebrow tipped with white: cheeks black; throat, side of neck, and chin greenish.-black; sides of upper neck streaked with white or drab; some of the feathers of mantle drab externally; under surface pale vinaceous grey, more buffish on abdomen; thighs blackish externally, internally white; under tail-coverts, under wing-covers and axillaries white; flights below blackish, with white internally; bill red at base, yellow at tip; feet yellowish; irides brown; naked orbital skin orange-yellow. Female smaller than male

Jerdon says (*Birds of India,*) :

The Pied starling is more abundant in the northern Circars than anywhere else where I have seen it. It associates in vast flocks of many hundreds, feeding among cattle. In general it is only found in small parties. It feeds, like the others, on grain, fruit, and insects. It is a familiar bird, feeding close to houses, and breeding on trees near houses-sometimes -- as at Sangor, in the midst of the town; though, as Mr. Blyth says: It does not venture into the streets in Calcutta.

It makes a large nest of sticks, grasses, and , feathers, usually about eight or ten feet from the ground, and lays three or four eggs of a clear greenish- blue. It breeds from April to June or July, according to the locality. It is very often taken young, and laid; has a pleaant song, and is a great imitator of other birds.

This handsome Starling reached the London Zoological Gardens in 1871. It did not live exactly at peace with its associates, " but is not spiteful if they leave it alone, but, if another bird repeatedly comes unpleasantly close, he stretches his long bill far up to keep it away, and exhibits an extremely extraordinary aspect. When he is anxious or distressed one hears him utter clear whistled notes. Finally, the Pied Starling is a musical bird, and its song is altogether the best Starling song that I know of.

3. Andaman Mynah *(Sturnus erythropygius adamamensis)*
This is also treated as a Starling. Butler describes it as follows:

> Above ashy-grey, whiter on the rump; scapular edged with white; wings black; the lesser-coverts edged with slate grey ; flights glossy greenish externally ; tail black, glossed with green, all the feathers excepting the central ones with a white marking, increasing in size outwardly, at end of inner web, the outermost being half white; head and neck all round and under surface white ; vent and under tail-covert tinted with fawn; flights below dusky with ashy inner web; bill and feet bright waxy yellow; base of lower mandible and tomium bluish-slate ; irides pearl blue. with bluish naked orbital ring. Female smaller and with shorter wing.
>
> Hab., Andaman and Nicobar Islands.
>
> I have not discovered any field-notes on this species, but it is now a well-known show bird. The London Gardens purchased a specimen in January, 1885.

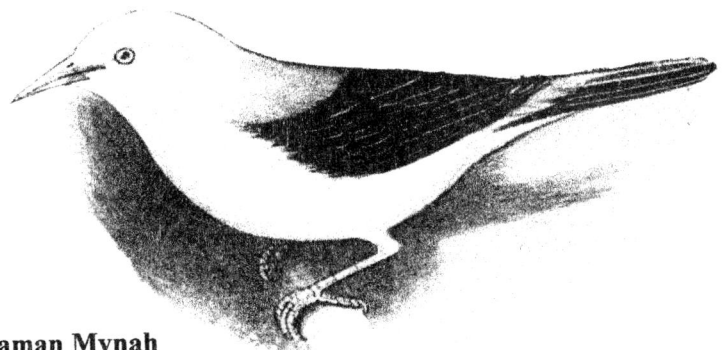

Andaman Mynah
Also known as Andaman Starling

4. Pagoda Mynah *(Sturnus pagodarum).*

Again Butler (ibid) has described the species:

> Above pearl-grey; wings, except covert, blackish, edged: with white near the shoulder; secondaries more or less grey; tall dusky grey-black, tipped with whitish; crown and crest greenish-black; ear-coverts and sides of head buff with paler streaks ; under surface cinnamon-buff ; bill slate-blue at base, greenish in centre, yellow at tip; feet bright yellow; irides greenish-white.
> Female similar, but smaller, and with shorter crest.

Hab. , —Afghanistan, India generally, and Ceylon.

Jerdon says *(Birds of India)* :
At Madras it feeds chiefly on the ground, among cattle, picking up grasshoppers and other insects. It also feeds on trees on various fruits, berries, and flower-buds, and occasionally insects. Adams says that in Cashmere it feeds on the seeds and buds of pines. When the silk cotton tree comes into bloom, is always to be found feeding on the insects that harbour in the flowers, I observed this at Jalna, and Blyth remarked the same at Calcutta. At Madras, it breeds about large buildings, pagodas, houses, etc. and lays three or four greenish-blue eggs. Mr. Philipps records it as building in holes of trees. It has a variety of calls, and a rather pleasant song. It is frequently caged and domesticated, is docile and hardy, and will imitate any other bird placed near it. Like the others of its tribe it is lively in its manners and actions, and has a steady swift flight.

It reacheds Amsterdam Zoo in the year 1853. It is, of course, a well-known show bird; a specimen was purchased by London Zoological Society in 1893, and possibly the first breeding was in September, 1901 -- Mr. Farrar recorded his success in breeding the species in an outdoor aviary; the nest was formed of small sticks in a box; the cock and hen incubated alternately about thirteen days; three young were reared. The eggs are described as small and blue like a Starling's.

5. Rothchild's Grackle *(Leucospar rothchildi)*

This has been described in *Encyclopaedia of Aviculture,* A Rutgers and others, as follows:

> The heavy body, pleasing proportions, and the thick graceful crest make this the most beatiful of the mynahs. The marvellous crest is composed of many slender 1 in. (2.5 cm) long; supple white feathers, starting at forehead and extending back to lower nape. Those on back of head and upper nape are particularly long. Colouring snow-white with black on a broad tip of tail and on ends of flight feathers. Bare facial mask of blue surrounds eyes and tapers to a point on sides of neck. Thick beak bluish-grey with dull yellowish-horn on tip. Feet and legs bluish-grey. Length : about 10 in. (25.4 cm). Sexes alike, except that females are somewhat smaller and feathers of crest are about 1 inch. (2.5 cm) long.

Apparently they eat the fruit of the melon tree flying around in groups of 20 to 30. The species was discovered in 1912 and A Ezra, a famous bird keeper, obtained 5 specimens and made many attempts to breed with them. He discovered that they were quite difficult to breed with. Even in the wild they are quite scarce.

Rothschild's Mynah
One of the most beautiful of Mynahs, but difficult to breed in captivity.
(After a Dutch painting)
Note the long, white crest.

6. Common Mynah *(Acridotheres t. tristis)*

This has been decribed by Butler (ibid) as follows:

> The prevailing colour is vinous brown, deeper, richer, and more glossy above than below; the crown, nape, lores, ear-coverts, and sides of face are glossy greenish black; the feathers on the forehead erected, but hardly forming a crest; eyelids black, but the naked patch below and behind the eye bright ochre-yellow ; the chin, throat, and breast black, less glossy the the upper parts, and seeming almost ashy in certain lights ; bastard-wing black, some of the feathers white externally ; primary coverts white; primaries black, white at the base, and brownish internally, the inner secondaries blackish, but the outer ones deep glossy brown; tail dull black, the central feathers slightly greenish, the remainder tipped with white, which increases in width from within outwards; centre of abdomen brownish white; vent and under tail-coverts pure white; bill and feet ochre-yellow, the claws browner, iris chestnut brown.
>
> The female is very like the male, but the bill appears to be slightly longer and the wings are shorter.
>
> Hab., Afghanistan, India. generally; Burma., and Tenassenm ; introduced into Mauritius.

The common Mynah is not specially striking in colour; it is about the size of a Blackbird. It is about 10 inches (25 cm) and is one of the commonest birds in the country, affecting towns, and villages. It roosts generally in large numbers, in some particular tree in a village or town and morning and evening keeps up a noisy chattering. Soon after sunrise the birds disperse and in parties of two, four, six. or more, wing their way in dIfferent directions to their various feeding-grounds. Some remain about villages looking out for any fragments of cooked rice that may be thrown out by the side of a house or even coming into a verandah for that purpose. Others attend flocks of cattle which they follow while grazing, picking up the grass hoppers disturbed by their feet, while

MYNAH BIRDS

some hunt for grain or fruit. It has a great variety of notes, some of them
pleasing and musical, others harsh; some have a resonant metallic sound. This bird breed like our English Starling, in nooks and under eaves of houses, or in holes in trees; it lays four or five pale bluish-green eggs,

In captivity the Common Mynah is said to become very tame, and to learn both words and sentences. Incubation is 15 days being incubated by the female and the male helps with the feeding.

Common Mynah

7. Brown Mynah *(Acridotheres f. fuscus)*

Butler (ibid) describes as follows:

> Above dull slate-grey, clearer on hind neck and mantle; lesser and median coverts bronze brown with greyish margins; greater coverts and inner secondaries with, black margins; primary-coverts white; remaining wing-feathers black, the primaries white at base and bronze brown at end of inner web; outer secondaries bronzy externally; tail black, tipped with white, greenish on outer webs of feathers; crown and sides of head greenish black; throat and chest dark slate-grey, shading into ashy buffish on breast, sides and flanks; abdomen clear buffish; under tail-coverts creamy white; thighs dark slate-grey; under wing-coverts blackish, tipped with grey; axillaries ashy buffish; flights below blackish; a white patch at base of primaries; bill blue-black at base, orange-yellow at tip; mouth bluish; feet orange-yellow; claws greenish horn; irides bright yellow.
>
> Female similar. but with shorter wings.
>
> Size 9 inches (22.9 cm).

There is also a Southern Brown Mynah which is:
> *Larger; bill orange-yellow, dusky on sides at base; feet yellow; irides bluish grey.*

The typical form inhabits the Sub-Himalayan region to the Central Provinces of India and eastward to Assam, Burma and Tenasserin; and its southern race, Southern India, as high as the Godavery Valley on the east and the neighbourhood of Ahmedabad on the west. (Sharpe.)

Jerdon observes (*Birds of India*):
This bird has almost the same habits as the common Mynah, like it often attending cattle, but also frequently seen in gardens, as at Ootacamund, eating seeds and fruit of various kinds; and it is very often seen clinging to the tall stem of the large Lobelia, so common on

the Neilgherry hills, feeding on the small insects (bugs chiefly) that infest the capsules of that plant. It is most abundant on the Neilgherries, where it is a permanent resident, breeding in holes in trees, making a large nest of moss and feathers, and laying three to five eggs of a pale greenish-blue colour.

In behaviour it is similar to the Common Mynah.

8. Bank Mynah *(Acridotheres ginginianus)*

Alec Brooksbank, in *Foreign Birds for Garden Aviaries* had experience of the Bank Mynah and decribes it as follows:

> The Bank Mynah (A. ginginianus) is not so big as theCommon Mynah, being only about 9 inches long and the body is slate grey with putty buff in the centre of the abdomen. The wings and tail have buff marks on them, bill deep yellow, wattle red. This bird is not so partial to human habitations and lives more in the fields and countryside and is particularly addicted to the riversides where it lives in colonies and builds exclusively in earthen banks or cliffs.

Bank Mynahs

9. Chinese Crested Mynah *(Acridotheres cristatellus)*
The description is as follows (Butler):

> The adult bird is silky blue-black, with a some\what. irregular crest of re curved feathers from the middle to the base of the bill {but no crest on the crown as represented in some scientific works). One bastard wing-feather is white towards the end of the outer web; the outer half of the primary coverts, the basal half of the primaries, and the base of the inner web of the secondaries are white ; the tailfeathers are tipped with white. Bill pale yellow, the base of the lower mandible pink, feet orange, iris deep amber yellow. Young birds are brown and have no crest; the bill and feet are browner, and the iris pale greenish- yellow.
>
> Hab. , Central and Southern China, and the Island of Luzon (Philippines), supposed to have been introduced.

It has been observed that this common resident breeds in holes, in trees and walls, as well as under the eaves or houses. The nest is a regular rubbish-heap of dry grass, straw, leaves, feathers, etc. The wing and tail feathers of pigeons, kites, crows and magpies are largely used. In every nest examined by Rickett there was a snake's slough or part of one, and our men were once told by a native that every Mynah's nest was thus provided.

The eggs are pale greenish blue. These birds are very noisy and pugnacious in spring. Although it is one of the most abundant and most charming of Starlings in the trade it is nevertheless unsociable; as well as spiteful, violent. and easily excited. Song copious and pleasing; it also imitates the notes of other birds, learns to speak excellently; is unusually tame. It is fond of berries and other fruits as well as grain.

This species is quite large (10.50 inches) and they have been known to be very entertaining, whistling, imitating speech, bowing and generally showing off. They appear to have quite a good life span, possibly in excess of the normal 20 years.

Chinese Crested Mynah
Drawn from life by Dr Butler

10. Golden Crested Grackle (Ampeliceps *coronatus*)
Described as follows:*

> About eight and a half inches in total length, the Golden Crested Mynah is one of the most colourful of the Mynahs with its contrasting glossy black and golden yellow. A far eastern species, it is to be found in the Malay Peninsular, and westward through Thailand to the Tenasserim district of Burma.
>
> The sexes are dissimilar, the adult male having a clear yellow crown, face, chin and throat, together with a long yellow crest, lying flat on the head. Further, there is a yellow wing patch, and bare skin around the eye is also yellow. The bill is orange yellow, the legs and feet being a duller shade. The remainder of the plumage is a glossy jet black. The female basically is the same colour but lacks the throat patch of yellow, the crest is not as long and neither is the yellow of the head as extensive.

A number of these birds were received by Mr. Boehm direct from Burma in April, 1962, from which a pair were selected for retention, the balance being given to various zoos. In the August of the same year a planted aviary, heated for the winter, was prepared for them, and this they shared with a pair each of Striped Kingfishers and Natal Robins. The diet of all three species was very similar; that of the Mynahs being fresh chopped fruit and blueberries, ground raw beef, a soft food mixture, and live food in the form of meal worms and crickets.

In addition to tree stumps with natural nesting holes, a covered nest box, eighteen inches high by nine inches square, was fixed to the framework of the aviary at about eight feet above ground level. It was provided with a two and a half inch entrance hole and had a sloping, hinged roof so that we could examine the interior as and when considered necessary.

---

* An article *Breeding the Golden Crested Mynah*, Chas. Everitt, NJ, USA, in *Foreign Birds*, Foreign Bird League magazine, July/August, 1964

It was not until May, 1963, that they showed any signs of breeding, and, at that time, the female was seen carrying twigs, rootlets and fresh leaves into the nest. She laid her first egg on May 10th, the final clutch being three. Althought these all turned out to be infertile, they did provide material for the following data. Averaging 32 by 19 mm., they were pale blue, two of them bearing faint brown markings and thread-like lines, mainly at the thicker end. The third egg was completely unmarked. Another round of three eggs was started on June 24th, these being identical to the first three, even to the extent of one egg being clear pale blue. Incubation, seemingly by the female only, and beginning with the laying of the first egg, lasted for fourteen days. Bare nestling down, the newly hatched chick had a yellow gape, margined in orange yellow. The other two eggs were peeping the following day and marks could be seen where the egg tooth was starting to break through. However, both eggs were ejected from the next box prior to either chick cutting its way out. The rearing of the solitary nestling was under-taken by both parents, live food being the diet for the early days. From about the tenth day onwards the male was seen to take pieces of fruit and raw beef into the nest, the duration of his stay in there indicating that it was fed to the nestling.

The baby Mynah's eyes were open at six days old but, except for dark markings under the skin, there was no sign of any feathering. Finally, at about the eleventh day, quill feathers began to show through and the general feathering up followed at a steady pattern from then on. At eighteen days old, yellow was showing on the head, together with pale yellow markings in the wings.

Vacating the nest at twenty-three days, it was seen that the crown of the head was yellow, as was under the chin, and that the primaries had a pale yellow bar across their centre. Both parents continued to feed the fledgling which attained independence at thirty-five days old. All three birds were left together for a further six months, by which time the young bird had acquired the markings of a male. At about this time the father began to chase it around so it was removed to one of the stock pens in the main bird house.

A month later, at the time of writing this, the female is sitting on another two eggs. As it has been stated in certain books that the young have all black heads, a fact disproved in the breeding described above, at least so far as young males are concerned, it is hoped that a female may he reared at some future nesting so as to ascertain if such remarks apply only to that sex, as young birds.

Coleto Mynah (after a Dutch painting)

11. Coleto Mynah *(Sarcops c. calvus)*

Butler (ibid) has described the species as follows:

> Dark glossy cinereous, blacker at base of feathers, the back some times mostly brown or black; a white patch on upper part of scapulars ; wings and tail black; head naked, dull pinky-white or flesh-red, excepting the lores, forehead, a line down centre of crown, joining a collar which passes round the ear-coverts, which are also of the same colour, the cheeks and the under surface, all of which are black; sides and flanks silver-grey; under tail-coverts washed with dark cinereous; flights below browner than above; bill and feet black, toes and claws brown; irides rufous-brown or chestnut. Female similar; but said to have, a longer wing.

Hab., Philippines and Sulu Islands.

It is also known as the Bald-Headed Starling.

Mr. J. Whitehead (*The Ibis*) stated:

Quite one of the ornithological features of the Philippines. Like the Great Hornbill, this species has also been noticed by the Spaniards, and is known to them as the **Collato**. It is supposed to learn to imitate the human voice, and for that reason it is often kept in a cage. The Collato is a busy, lively bird, being found in numbers of the forests when its favourite food is ripe. It is also very partial to dead tree trunks, nesting and roosting in the numerous Woodpeckers borings

The noise made by the wings during flight is very audible. In Samar a pair were very busy prospecting some old posts within a few feet of our house, but we left before: they had commenced to build. The note is a peculiar click, metallic but not displeasing.

12. Ceylon Mynah *(Gracula ptilogenys)*

Butler (ibid) gives a full description:
> Black, glossed with purple on the crown, sides of head, neck all round, hind neck, mantle, broad borders; of greater wing-coverts, and narrow margins to inner-most secondaries, tail, throat, chest, thigh, edges of under wing-coverts and axillaries towards edge of wing; glossed with green on back, rump, upper tail-coverts, scapulars, wing-coverts, foreneck, breast, and remaining underparts, ; primaries with the usual white patch; wattles apparentlly wanting, lappets bright yellow; bill orange-red:, the upper mandible black from gape to nostril. and the lower for nearly half length; feet gamboge yellow, claws blackish; irides greyish-white dappled with brown. Female with weaker bill ; irides white or yellowish-white.
>
> Hab., Sri Lanka

This bird frequents the top of tall trees and cliffs and flies from great heights when it utters a shrill cry which is metallic in sound. It also whistles and makes guttural calls, described as choooke, chi-ooope, when perched. They tend to live in the tops of the trees. They eat fruits of all kinds, especially nutmegs and wild cinnamons.

In the wild, it breeds in June, July and August, nesting in a hole in a tree, usually previously used by a woodpecker or other bird. The eggs are laid on the bare floor of the cavity. They are oval in shape and are a pale greenish-blue; size 1.30 inches x 0.96 inches in breadth. The birds are about 9.50 inches (24cm

In captivity they fare fairly well, but must be given a strict diet of milk soaked bread, mealworms, pieces of fruit and some form of insectivorous mixture. Apparently, if the correct diet is not given they will not survive long. In cages they will learn to talk, but are probably better in a bird room and aviary because their liquid droppings can cause problems if kept indoors.

MYNAH BIRDS

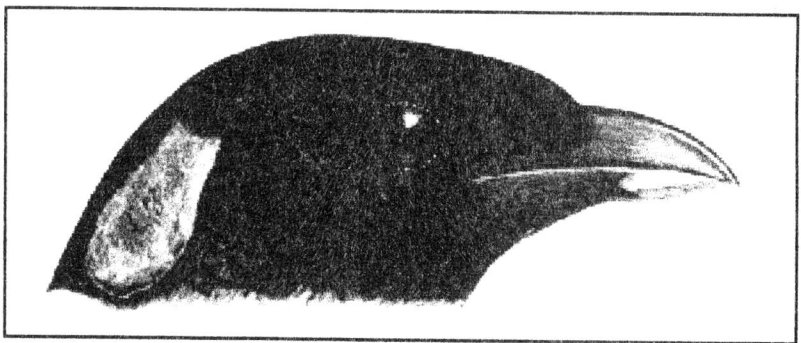

Head of Ceylon Mynah

Head of Greater Hill Mynah

Used to be called the Malay Crackle (Eulabes javanensis). It would seem that Dr Butler confused this with another species because the markings he gives in *Foreign Birds for Cage & Aviary* **are not correct. This must have been a different, related species; more likely the Indian Mynah** *(Gracula r. intermedia).*

13. Lesser Indian Hill Mynah *(Gracula religiosa indica)*. Also known as the Southern Hill Mynah, it comes from Southern India and Sri Lanka. It is about 9.50 inches (24 cm). They have been bred in captivity, but are not as able as the other species in this category.

14. Indian Mynah *(Gracula religiosa intermedia)*. From India, Burma, southern China and the hills of the southern Himalayas. Size: around 11 inches (28 cm).

15. Greater Hill Mynah (*(Gracula r. religiosa)*. From Java, Bali, Malaysia and Sumatra and adjacent islands. Size 12 inches (30cm).

These are similar and related. They are largely black, but with differences in size, head markings and slight plumage variations.

THE TALKING MYNAHS

The species mentioned on this page are the main ones for keeping and, hopefully for training as Talking Birds because this is the main attraction for keeping them. These were described by H A Fooks as follows*:

> The Talking Mynahs, or Grackles are resident over most of the afforested parts of what was once known as the Indian Empire, Malaya, Burma, Siam, Annam, Cochin China, and the Andaman Islands.
>
> Their plumage is almost entirely a glossy black, with a very distinctive pair of fleshly wattles on either side of the nape, and usually two patches of bare skin below the eye. Their bills are powerful and curved, with weak rictal bristles round the base. The feathers on the crown are short and curly, forming a parting down the centre. The feet are very powerful, and unsuitable for walking on the ground. The bird progresses terrestrially in hops, but never looks really comfortable whilst doing this.

Foreign Birds, June, 1957

MYNAH BIRDS

Greater Hill Mynah
(Gracula religiosa religiosa)

Indian Mynah *(Gracula religiosa intermedia)*

A NATURALIST IN BURMA

B E Smythies spent his working life in the forests of Burma and Borneo and part of his work is reproduced here, thus giving a modern perspective on the Mynahs and some Starlings. (***Birds of Burma,*** 3rd edition, 1986)

The Hill Mynah is resident in the better wooded parts of the Oriental Region; the whole plumage is glossy black, but the most distinctive feature is the presence of fleshy wattles on the nape, usually accompanied by bare skin on the sides of the head. They have an immense repertoire of notes, some of which are melodious and others noisy and raucous; they are also admirable mimics, copying the notes of other birds freely in a wild state and learning to talk well in captivity.

The Glossy Starling ranges from NE India through Burma and Malaysia to Australia. The plumage is intensely glossy and the wing long and pointed. In habits they closely resemble some of the mynahs.

The starlings and mynahs include some of the commonest and most familiar birds in Burma.

HILL MYNAH, *Gracula religiosa* Linnaeus

SUBSPECIES: *intermedia* A. Hay, *religiosa* Linnaeus

LOCAL NAME. Burmese: tha-li-ga.

Plumage black except for a white wing-patch. An arboreal forest bird of the foothills.

VOICE. Described above. Perhaps the most distinctive note is a high-pitched Sibilant whistle of great power.

HABITS AND FOOD. The Hill Mynah, or Talking Mynah, is a sociable bird and is usually seen in large parties in the tree-tops; it is a characteristic bird of the teak forests of the Pegu Yomas, where its variety concert is a frequent source of entertainment. Owing to its vocal prowess it is a favourite cage bird in India, and there used to be a considerable export trade in these birds from Rangoon, several thousand being shipped at a time under the name of Hill Mynahs; owing to the cruelty involved an attempt was made to suppress the

trade by protecting the birds under *section 9* of the **Wild Life Protection Act.**
NEST AND EGGS. Most eggs are laid in April and May, but Hopwood took them in Arakan in June; they are laid in a hole in a tree and are a bright deep blue, blotched and spotted to a varying degree with chocolate and reddish-brown andwith underlying marks of pale blue, the colour fading very quickly after the eggs have been laid; there are usually 2 or 3 in a clutch.
STATUS AND DISTRIBUTION. Throughout most of the Oriental Region. It is well distributed throughout Burma in the forests of the foothills, but is not often seen in the higher hills, although Wickham described it as found in all the upper Burma hills and a single bird has been observed in North-East Burma at 6,800 feet.

PHILIPPINE GLOSSY STARLING , *Aplonis panayensis*
SUBSPECIES: ***affinis*** (Blyth), strigatus (Horsfield)
IDENTIFICATION. 8". Whole plumage black glossed with brilliant green. It has a sharp metallic single note and is found in large noisy flocks in the tree-tops in open well-wooded country; in Tenasserim it roosts in coconut palms and pagoda *htees.*
NEST AND EGGS. Hopwood found many colonies breeding round Mergui from March to June and Christison found it breeding in Arakan in May. The nest may be a hole in a tree, a bridge, the roof of a building, or in a toddy palm at the junction of a leaf with the trunk; it is a round cup of roots, grass and leaves. The eggs, usually 3 in number, are miniature Hill Mynah's eggs.
STATUS AND DISTRIBUTION. Indo-chinese subregion and Malaysia. Common in Tenasserim, very local in Arakan, extending to Assam but there are no authentic records from any other part of Burma.

SPOT-WINGED STARLING, *Saroglossa spiloptera spiloptera* (Vigors)
IDENTIFICATION. 7 1/2". Sexes differ. The male has black edges to the feathers of the cap and mantle, black sides to the head, a white wing-spot and rufous under- parts, darker on the chin and throat. The female is browner above and grey-brown below with white edges to the feathers. Found in flocks in the

tree-tops; often in *letpan* trees.

STATUS AND DISTRIBUTION. Breeds in the Himalayas from Nepal westwards. It seems to be an erratic winter visitor to Burma and has been recorded from North-East Burma, the plains of Southern Burma as far south as Elephant Point, Southern Shan States, and Karen Hills, all the records being from December to February.

COMMON STARLING, *Sturnus vulgaris* **Linnaeus**
SUBSPECIES: *poltaratskyi* Finsch
IDENTIFICATION. 8 1/2". Plumage black with a purple gloss on the head and foreneck and a green gloss on the upper-parts.
STATUS AND DISTRIBUTION. A Palaearctic species. A rare winter visitor to North-East Burma, one having been obtained at Fort Hertz and two seen near Myitkyina.

WHITE-CHEEKED STARLING, *Sturnus cineraceus* (Temminck)
IDENTIFICATION. 9 1/2". Crown, nape and sides of head black; ear-coverts white streaked with black; rest of plumage ashy-brown or dark ashy except for a white band across the rump and the creamy-white abdomen and under tail-coverts.
STATUS AND DISTRIBUTION. Breeds in E Siberia and Japan and winters in the E Oriental Region. A rare winter visitor to N Burma, having been obtained at Kamaing and Bhamo. There is one record from Yunnan.

WHITE-SHOULDERED STARLING, *Sturnus sinensis* (Gmelin)
IDENTIFICATION. 8". General colour white to pale grey, ashy on the neck and back; tail black broadly tipped with rusty; primaries black with metallic green gloss.
STATUS AND DISTRIBUTION. Breeds in E Siberia and N China, and winters in the E Oriental Region. A rare vagrant to Burma, recorded only from near Pegu, but likely to occur in Tenasserim.

MYNAH BIRDS

CHESTNUT-TAILED STARLING, *Sturnus malabaricus* (Gmelin)
SUBSPECIES: *nemoricola* (Jerdon), *malabaricus* (Gmelin)
A small pale grey starling liable to be confused with the Vinous-breasted Starling, but distinguished by smaller size, colour of bill, chestnut under tail-coverts and lighter grey wings.
VOICE. It has querulous, rather high-pitched call-notes and a quite pleasant song.
HABITS AND FOOD. This is the most arboreal and active of the starlings, found in flocks in the tree-tops in open forest and in gardens, small groups of mango trees scattered about the paddy plains, flowering letpan trees, etc. On several occasions in the cold weather they have been observed late in the evening or just after dawn huddled close together in rows, like Ashy Wood-Swallows, on the bare branches of trees standing out in the open.
NEST AND EGGS. Breeds from April to June. The eggs are laid in hollow trees, including palms, and are 3 to 5 in number and pale green in colour without markings.
STATUS AND DISTRIBUTION. Widespread in the Oriental Region, and common throughout the plains and foothills of Burma outside deep forest. The sub-species malabaricus was collected at Homalin by Milton.

PURPLE-BACKED STARLING, *Sturnus sturninus* (Pallas)
IDENTIFICATION. 7 1/2". Crown, nape, neck and upper back grey with a metallic purple patch on the nape; back, rump and wing-coverts metallic purple; tail metallic green; a ring round the eye white; sides of head and under-parts grey. Said to resemble the preceding species in habits.
STATUS AND DISTRIBUTION. Breeds in E Siberia and N China, and winters in the E Oriental Region, including Tenasserim (extremely abundant in April round Tavoy); there is one record from Pegu.

BLACK-COLLARED STARLING, *Sturnus nigricollis* (von Paykull)
The white rump is conspicuous in flight. A familiar starling of E Burma.
VOICE. A great variety of notes and phrases of which *chirr-chu-chu and chirr-*

ta-chu-ta-chu are the commonest; also a hoarse jay-like *kraak-kraak*.

HABITS AND FOOD. This large starling will be familiar to anyone who has ever been to Maymyo, where it waddles about on the lawns and open grassy spaces picking up grasshoppers and crickets. It resembles the Common Mynah in habits and spends most of its time on the ground, flying off in the evening to a common roost.

NEST AND EGGs. The breeding season is from April to June, with second broods into August. It builds a huge, domed untidy nest of grass, leaves, and odds and ends, high up in a tree; several birds sometimes breed in company, and the same site may be used annually. The eggs, 3 to 5 in number, are larger editions of the eggs of the Common Mynah.

STATUS AND DISTRIBUTION. Widespread in the E Oriental Region. A common plains bird in N Burma, and N Central Burma (where the Chindwin river seems to be its western boundary). It is found west of the Chindwin further north, at Htinzin north-west of Mawlaik (Baillie). It extends through the Shan States to Tenasserim (where it is rare), but Oates mentions some specimens from the Tenasserim river.

VINOUS-BREASTED STARLING, *Sturnus burmannicus* Jerdon

SUBSPECIES: *burmannicus*, Jerdon, *leucocephalus* (Giglioli & Salvadori)

Whole head and breast dirty white, upper-parts dark grey; under-parts purplish; the plumage varies greatly with the seasons, but the red-tipped bill is usually noticeable.

VOICE. It has a chattering note, which parties constantly utter while feeding.

HABITS AND FOOD. A typical dry zone bird, usually seen in large parties feeding on the ground in open grassland, gardens, cultivation, and scrub-jungle. It has the family habit of collecting in a communal roost at night; the roost may be in reeds, sugar-cane, bamboo clumps, or similar cover.

NEST AND EGGS. Breeds from April to June, with second broods up to August. It builds an untidy nest under the eaves of houses or in holes in trees and the roofs of thatched huts.

STATUS AND DISTRIBUTION. Widespread in the E Oriental Region, not west of Burma. Its stronghold is the dry zone of Central Burma, and it is the only starling or mynah (apart from the Common Mynah round villages) normally seen in the uplands west of Meiktila. Northwards it extends to Myitkyina, where it breeds, but probably not much beyond, and southwards it is a cold weather visitor to the out-skirts of Rangoon. Beyond that again we find it (subspecies *leucocephalus*) resident in Tenasserim. It is not found in the higher hills of N Burma, but is common on the Shan plateau.

ASIAN PIED STARLING, *Sturnus contra* Linnaeus

SUBSPECIES: *contra* Linnaeus, *superciliaris* (Blyth), *sordidus* Ripley, *floweri* (Sharpe)

IDENTIFICATION. A conspicuously pied starling with a distinctive bill. Common in the plains.

VOICE. No details of its call-notes seem to have been recorded.

HABITS AND FOOD. More rural and less urban than the Common Mynah; it is a bird of open cultivation, where it lives in small parties that spend their time hunting over paddy fields and grassland, and is much in evidence from the train window in lower Burma. During the breeding season family parties are commonly seen in the environs of Rangoon. Like the mynahs it roosts in huge, vociferous mobs.

NEST AND EGGS. Similar to those of the mynahs. The nest is usually built in a tree out in the open fields; kokko trees along bunded roads are a favourite site.

STATUS AND DISTRIBUTION. Widespread in the Oriental Region, and a common resident throughout the plains of Burma as far south as Mergui, but it is not found in the higher hills, except perhaps in the Shan States, nor in forest.

GOLDEN-CRESTED MYNAH, *Mino coronatus* (Blyth) (=*Ampeliceps coronatus*)

IDENTIFICATION. 8 1/2". A remarkable mynah, easily recognized by its bright black and yellow plumage, long crest, bare yellow skin round the eye and orange bill; young birds have the whole head black. An arboreal bird of the plains and foothills. A bird of dry open forest, usually in small parties, occasionally in pairs, very similar in voice and habits to the Hill Mynah, but it also utters a harsh metallic note like ***Crypsirina temia*** (Davison).

NEST AND EGGS. Have been taken in Tenasserim in April.

STATUS AND DISTRIBUTION. Indo-chinese subregion. In Tenasserim it seems to ; be not uncommon as far south as Tavoy but elsewhere in Burma it is rare and its status is uncertain; it has been recorded from the Mogok foothills (at 500 feet), the Pegu Yomas, and Plains of Southern Burma.

COMMON MYNAH, *Acridotheres tristis tristis* (Linnaeus)

LOCAL NAME. Burmese: *zayet* (a name applied to all mynahs).

IDENTIFICATION. In flight the rounded white-tipped tail and the large white patch on the wing are conspicuous. Common plains species.

VOICE. A strange mixture of harsh gurglings and liquid notes, *keeky-keeky-keeky, churr-churr, kok-kok-kok,*. the last notes are invariably accompanied by a quaint, stiff bobbing of the head, generally close in front of the bird's companion. Other notes recorded are *che-wee -che-wer -chee-wee* when contented, also quite a cheery nondescript warble, and a harsh screech when worried; if disturbed when feeding the birds rise with a querulous note.

HABITS AND FOOD. This mynah shares with the Tree Sparrow and the House Crow the distinction of being the commonest and best-known bird in Burma. The generic name *Acridotheres*, meaning "the grasshopper hunter, " is unusually apt, for it strides about on the ground with rapid, determined paces and a slight waddle, peering under the fallen leaves in a comical manner; the specific name tristis, however, is somewhat of a misnomer, for the Common Mynah is a perky, self-confident bird, always sprightly and vivacious, and perhaps the only one (as Stanford observes) that habitually walks about singing while looking for food. It is also pugnacious, and I once watched a tremendous battle between two birds in a timber yard in Rangoon; how the fight started I do not know, but when I came upon them they were locked in mortal combat on the ground and so absorbed that they ignord my approach; they did not spar after the manner of fighting cocks, but practised infighting; after a fast and furious five minutes they separated and flew off, apparently none the worse, although it seemed inevitable that one of them would be left slain upon the field, so fierce was the encounter .

This mynah, more than any of the others, is linked to human habitations, and is rarely seen far from towns and villages. It is usually paired, and there is a very obvious affection between the two birds; they feed together and step occasionally to preen each other's feathers or to indulge in a few quaint remarks expressive of extreme self-satisfaction. They collect at night at some favourite roosting place, which may be a grove of trees or a sugar-cane plantation, and which they often share with other mynahs, House Crows, and parakeets; they only go to sleep after the most noisy and quarrelsome proceedings, and at intervals during the night short bursts of chattering are to be heard.

NEST AND EGGS. In the irrigated areas of Shwebo district "the earliest observation of nest building was on the 24th April. One nest in which young were being fed on the 16th July was being relined with sticks on the 1st August; on the 21st August the new brood was being fed with caterpillars, continuing to the 2nd September; on the 29th September fresh sticks were being taken in and by the 7th October the third clutch was apparently being brooded; by the 16th October the young were being fed, continuing to the 8th November. These dates give an incubation period of not more than 17 days and a fledgling period of at least 24 days" (Roseveare). The breeding season therefore extends at least from April to October. The nest is a shapeless mass of straw, twigs, feathers, etc. , stuffed under the eaves or in a hole in a wall, tree or well; the old nest of a kite, crow, or squirrel may be adapted and re-lined. The eggs, 3 to 6 in number, are blue in colour without markings.

STATUS AND DISTRIBUTION. Almost throughout India, Burma, and the Indochinese subregion. It is resident throughout Burma wherever there are towns or villages of any size; in the Kachin hills it is found in small numbers at outposts such as Sumprabum, Laukkaung, and Sadon, and is common in the Fort Hertz plain, but I have not seen it round Kachin villages, and the complete absence of all mynahs and starlings from the N gawchang valley, which appears suited to their requirements, is remarkable. In W Yunnan it goes up to 6,000 feet. Harington stated that it did not ascend the hills in Bhamo district, and I saw no signs of it at Sinlum Kaba; on the other hand, it breeds at Bernardmyo (5,000 feet) in the Mogok hills, and from there southwards is probably ubiquitous on the Shan plateau.

CRESTED MYNAH, *Acidotheres cristatellus cristatellus* (Linnaeus)
IDENTIFICATION. Resembles the next species, but is a larger bird with very thick frontal-plumes. Under tail-coverts black narrowly barred white.
STATUS AND DISTRIBUTION. Breeds in China and Yunnan, and in Burma has only been recorded from Malipa in the east of the Northern Shan States.

JUNGLE MYNAH, *Acridotheres fuscus* **(Wagler)**
SUBSPECIES: ? *fumidus* Ripley, *torquatus* Davison

WHITE-VENTED MYNAH, *Acridotheres javanicus* Cabanis
SUBSPECIES: *infuscatus* (Baker), *grandis* Moore
Authorities differ on the treatment of these forms. Deignan makes *torquatus* a subspecies of *mahrattensis* Sykes, which Salim Ali & Ripley make a subspecies of *fuscus*. Dean Amadon, 1956, thinks that *grandis* is probably a distinct species (A mer. Mus. Novitates No.1803: 32-34).
IDENTIFICATION. At a distance it can be confused with the Common Mynah, but is distinguished by darker plumage, absence of yellow face-wattle, and by the tuft of feathers above the nostril. In flight the white wing-patches and white-tipped tail are noticeable. The Jungle Mynah has a deep blue base to the yellow bill; yellow eyes; dark greyish-brown upper-parts. The White-vented Mynah has a longer crest, all yellow bill, and black upper-parts.
VOICE. No one seems to have bothered to record any notes on the voice of this common bird.

HABITS AND FOOD. In flight, habits, gait, and behaviour it resembles the Common Mynah and is often mistaken for that bird, though it is neither so bold nor such a scavenger. It is not a bird of high forest, as its name might lead one to suppose, but may be found in clearings or in open park-like forest; it is more typically a bird of open country and is frequently seen with cattle and buffaloes, often perched on their heads. It has been observed feeding on the young buds of a Ficus, on flowering *letpan* trees, and round the borders of a *jheel* where kaing grass was mixed with bushes and a few trees. In the breeding season it is a familiar bird in the environs of Rangoon.

NEST AND EGGS. Harington described mixed colonies of Jungle and Collared Mynahs and bee-eaters breeding in holes in steep, sandy banks along the Chindwin above Kindat; he was unable to decide whether the holes were originally made by other birds and then enlarged by the mynahs or excavated entirely by them. Pieces of snake skin were found in every nest he examined. Normally, however, this mynah breeds under the eaves or in holes in trees, etc. like other mynahs. In Myitkyina they start on the 20th February.

STATUS AND DISTRIBUTION. Widespread in the Oriental Region and a common resident in the plains of Burma and on the Shan plateau. The distribution of the two species in Burma needs re-examination.

WHITE-COLLARD MYNAH, *Acridotheres albocinctus* Godwin-Austen & Walden

IDENTIFICATION. 9 1/2". Distinguished from the Jungle Mynah by a white collar on the hind-neck and a much smaller tuft above the nostril (very noticeable in a mixed flock) ; in winter the white collar consists merely of buff tips to the feathers but in spring is very prominent. In habitat, behaviour etc. , there seems to be no difference .

NEST AND EGGs. *See under the Jungle Mynah*. Large colonies have been noticed in the sandy banks of the Uyu and Chindwin rivers near Homalin (Smith).

STATUS AND DISTRIBUTION. Manipur, Burma, and W Yunnan. It is common in the Chin Hills, N Burma and the Shan States.

[NOTE. -The relationship between the White-collared and Jungle Mynahs is peculiar; though so similar, the two appear to behave as two distinct species; ii; almost every locality in which the White-collared Mynah is found the Jungle Mynah is also found, but they do not seem to interbreed.]

Scene in Burma

MYNAH BIRDS

STATUS AND DISTRIBUTION. Manipur, Burma, and W Yunnan. It is common in the Chin Hills, N Burma and the Shan States.

[NOTE. -The relationship between the White-collared and Jungle Mynahs is peculiar; though so similar, the two appear to behave as two distinct species; ii; almost every locality in which the White-collared Mynah is found the Jungle Mynah is also found, but they do not seem to interbreed.]

City Dwellers
Many Mynah Birds prefer to live in villages, towns or cities in their native lands.

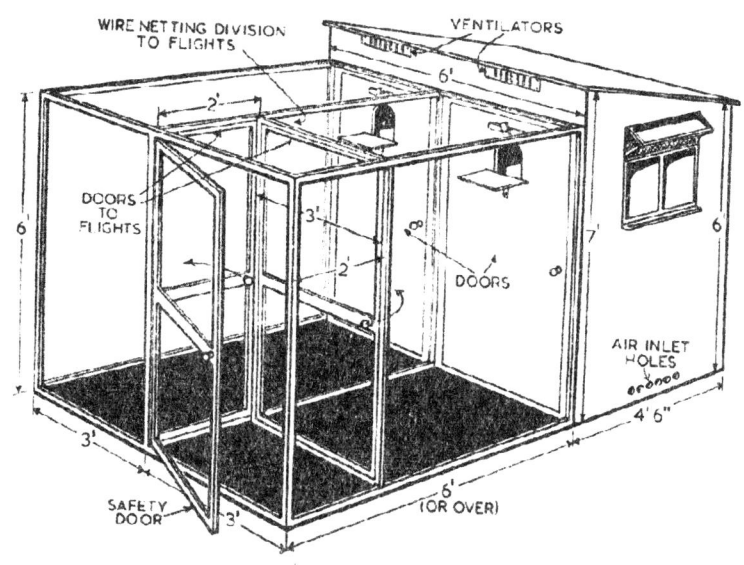

Simple Aviary
Designed for small parrot-like birds, but may be adapted.

3

AVIARIES
&
CAGES

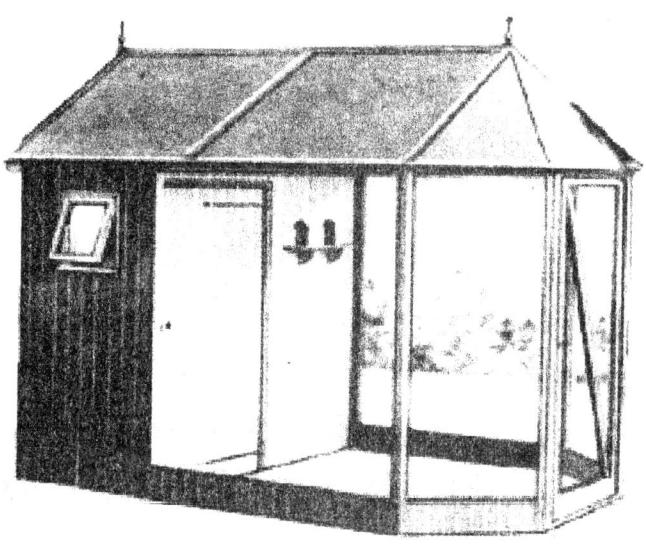

Shed with Covered Flight
Ideal for birds which require full shelter

Smaller Aviaries for a Pair of Mynahs
Right hand aviary has flight half covered and in winter shutters can be made for open section.

MYNAH BIRDS

PRINCIPLES OF BIRD KEEPING

The Mynahs have their own characteristics and come from different environments and conditions. Because of the diverse factors it becomes rather difficult to generalize too much, but there are principles which can be applied. Obviously the size of the species and the way it behaves in the wild is some guide to the way it should be kept, but this cannot be carried too far or no birds would be kept in captivity. Remember also that many are bred in captivity so will know no other way of being kept.

The guiding principles are as follows:

1. Compatibilty
Some Mynahs are quite aggressive, especially at breeding time and, although they are sometimes placed in a mixed collection, and some species are less aggressive than others, great care should be taken in placing them with others. If breeding, then keep the breeding pair in a separate aviary with adequate flying space.

2. Size of Aviary
Because many of the Mynahs are constantly flying there should be plenty of perches, suitably spaced, so that they can obtain the exercise to keep them fit. A shed (sleeping area) and aviary (flight) of around 2 metres x 2.5 metres is desirable, although a pair could manage with half that size. If birds are to breed they should have a reasonable amount of space. They should have plent of landing stages so they are able to fly from one to the other. In the wild they spend much of the time in trees and are not generally floor feeding birds.

3. Objectives
If the intention is to have a bird for talking then a suitable cage will be required. This enables the bird keeper to maintain contact with the

birds and the result should be that they become friends. Constant talking should result in words being imitated and the vocabulary will gradually extend. On the other hand, if birds are kept in cages the possibility of talking becomes rather restricted, although it can be very effective.

4. Suitable Cage Accommodation and Space

The Mynah bird with his ability to imitate and perform tricks is in great demand because he is much admired. However, they are not quiet trouble free birds such as canaries or budgerigars. At times they can be quite outrageous and try *tricks* such as spitting, hissing, and making strange sounds which are not acceptable.

The other feature is that they can be very dirty, throwing food around and splashing the cage and outside areas with very liquidy droppings. It means that a cage should be enclosed on three sides and the front should have a front shield of perspex about 6 inches (15cm) high so the droppings are kept inside. In addition, some bird keepers spread old newspapers which are removed daily and burnt.

If they are to be kept in a cage it would be better in some sort of bird room in the house or outside in a well insulated shed.

The use of shavings would be possible in an *outside* bird room because these can easily be removed when soiled and this idea is better than dealing with soiled newspapers.

The cage should be about 1 metre long and 0.50 m high and 0.40 m wide. The very large Greater Hill Mynah should have a cage about twice this size or some form of indoor aviary.

The talking Mynah is usually housed singly in a suitable cage. If left over the weekend make sure there is plenty of food and water. Although the conventional bird cage has a front door which is raised or lowered to open the preferred method with Mynahs is to have side doors so there is easy access.

Line the cage with a few layers of newspaper and remove one each day as it becomes soiled.

The open Parrot cage is not regarded as suitable because it is too open to the outside draughts. The larger wooden type which is normally used is more commodious.

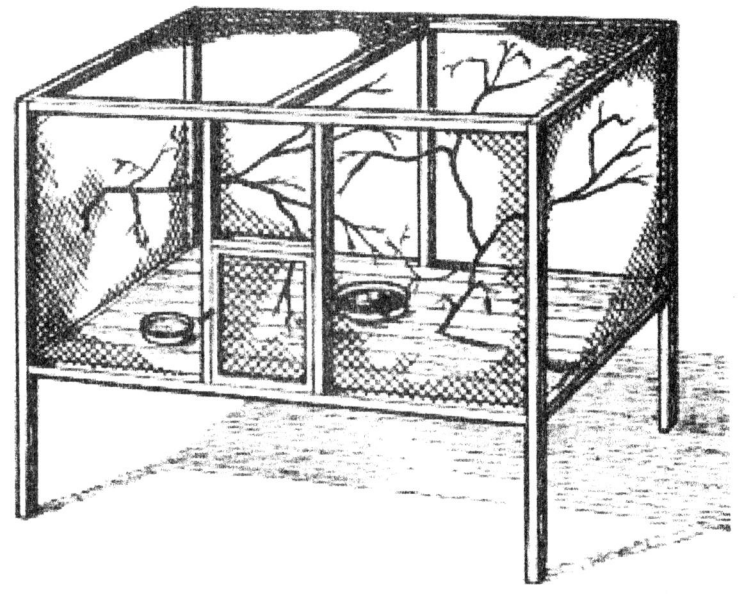

Small Aviary for Internal Use
Because of the loose droppings the sides would have to be protected or the outside would get rather dirty and splashed.

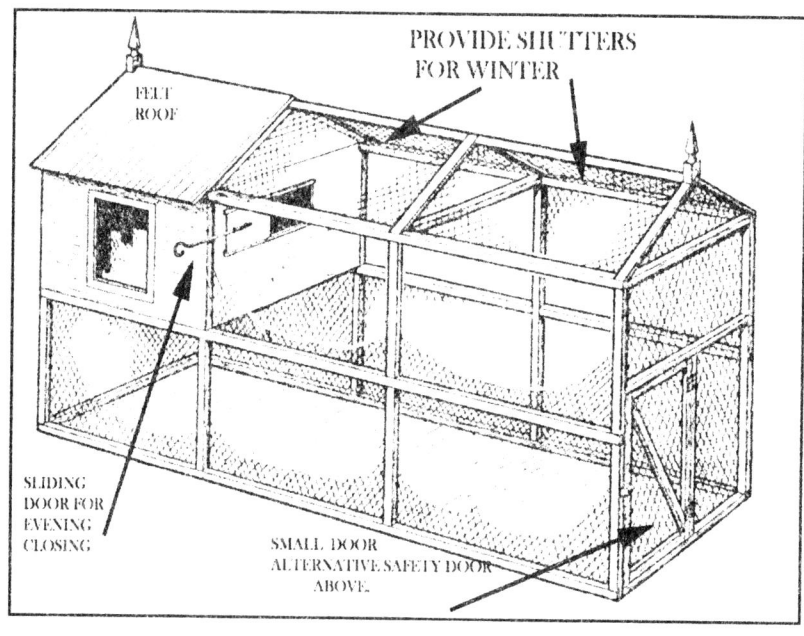

Above: A Basic Aviary and Sleeping Compartment
In Winter the aviary should be covered over with plastic covered frames or similar material so it does not become too cold. Usually the Mynahs will suffer if below freezing.

OPPOSITE
A landscaped aviary with ample room for Mynahs. Glass panels or perspex can be used to shut out the really cold weather in winter.

Large Aviary in Garden and suitably landscaped.

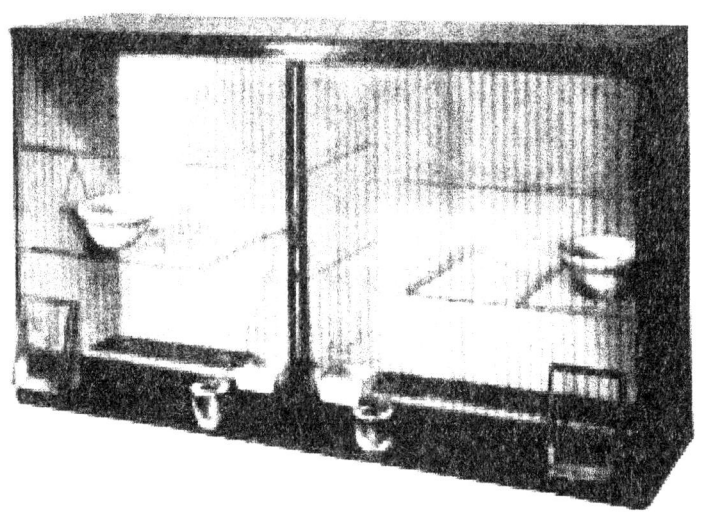

Double Breeding Cage with Partition

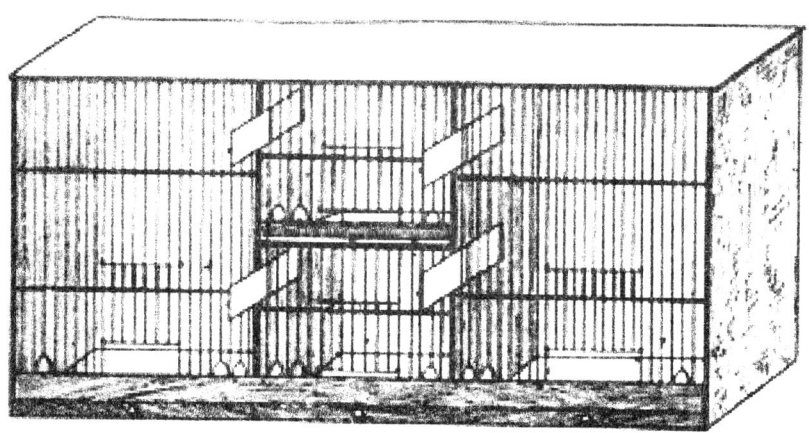

Cages for Keeping Mynah Bird
Home made type with partitions moved give over 1 metre exercise space.

BREEDING & SEXING

Many zoos or bird gardens have been successful in breeding some of the Mynahs, but not all.

Amongst these were Keston Foreign Bird Farm, and other aviculturists. One of the features seems to have been the use of very large aviaries where the birds could fly around freely.

This is not helped by the fact that male and female may look the same. Usually, in the case of the Hill Mynahs the cock bird will look bigger and the head is broader. If the pelvic bones are tight together then it is probably a male bird.

Sometimes the male has a deeper more masculine voice, but this is not always the case. In short, we can never be sure of the sex until breeding occurs, but either sex should be a good talker.

Features

ANDAMAN MYNAH
The wing of the female is much shorter than that of the male, but no difference of plumage is recorded.

MALABAR MYNAH.
The female is much paler than the male, and has yellower legs and whiter eyes.

BLYTH'S MYNAH
The female has a shorter wing than the male, is paler and duller; more rusty above, especially on the rump and upper tail-coverts; head more ashy, sides of face and throat greyer; the rufous feathers somewhat edged with ashy whitish; culmen of bill brownish; legs tinged with olive; iris greyer.

PAGODA MYNAH
The female is smaller than the male, and has a shorter crest, but no difference of plumage has been noted.

ROSE-COLOURED PASTOR.
The female is smaller than the male, has shorter wings, has a shorter crest, and is less brightly coloured.

COMMON MYNAH
The female has shorter wings than the male, but no difference of plumage has been indicated.

BROWN MYNAH
The female has shorter wings than the male, but appears to be similar in plumage.

CRESTED MYNAH
I have no doubt that the wings of the female are shorter than those of the male.

TRUE MYNAHS OR GRACKLES (MAINATUS, ETC.).
In these birds the bills of the males are stronger than those of the females.

BLACK-NECKED MYNAH
The female is a good deal smaller than the male, but no difference of plumage has been noted.

WATTLED MYNAH
In the female the wings are shorter than in the male: she is also browner, making the white rump more conspicuous; the upper tail-coverts brown; the wings are browner, the primary-coverts black; the head is feathered, not bare on the crown; the only bare parts are round the eyes, a yellow patch behind the latter, and the sides

GREATER HILL-MYNAH
The female is smaller than the male, but similar in colouring; her bill is considerably weaker.

MYNAH BIRDS

Get the Mynah Tame and Fearless.

In this way confidence is built up and the bird begins to show off and talks much better

TALKING

Teaching to talk should start at an early age. The first step is to handle the Gaper and get it tame. Start by saying **Hello Henry** or some other suitable name and then introduce new words once the bird has grasped at least one word.

Getting a bird very tame is important and, if possible, by inserting the hand in the cage, and holding titbits, it may be induced to perch on the hand or finger. If a little uncertain about the behaviour try wearing a glove, thus giving some protection.

Once the bird is able to repeat its name then introduce other single words and, once that is over, try short sentences. In this way the vocabulary will increase and the words will be repeated more clearly on the basis that practise make perfect.

Repetition is the main requirement and some bird-keepers have tried teaching by having a tape playing, with the sentences suitably spaced. Obviously, there is some skill required to organize the lessons. Moreover, the words should be uttered clearly and distinctly.

A CASE STUDY -- "MINO"

The Mynah bird, we are told, is closely allied to the starling family. It is an attractive creature, not so large as a pigeon, but far more graceful. His plumage is a lustrous greyish-brown, and he carries a pair of flat bright yellow wattles on his head. He has lovely eyes, and eye lashes as long and as heavy as those of the heroine of any modern novel. His bill is a bright yellow. This story is about a bird named 'Mino'.

He imitates the Banjo and speaks French and German.
When a New York reporter-- a lady-paid him a visit, he sat on the window-sill looking longingly out at the blue sky and the budding trees.

Then he turned to his owner Miss Thursby and said:

"Mamma, please let me out; I won't stay long,",

in the irresistible voice of a child.
Miss Thursby explained that it was unsafe to let him out. One day he was stolen, and we've never let him out in town since. Fortunately, the

hall boy saw a man pick him up and run away with him. He came up and told me, and my sister and I rushed out to hunt him. We found him in a saloon on Third Avenue entertaining a party of men.

'Have a glass of wine,' he would say to one, and.' Do take another,' to someone else. He's forgotten to say those things now since we are a temperance family."

When Miss Thursby, the owner, asked Mino to sing or play the banjo for her visitor, he saucily replied, " Go away."

Then Miss Thursby picked up a metal paper cutter and began to beat a tune on a big China vase. This was too much for the bird. He ducked his head, puffed out his feathers, and gave a perfect imitation of a banjo.

"Take your high notes," urged Miss Thursby, and up and down the scales he ran. It sounded for all the world as if the inside of his throat was strung like a banjo. A parrot will imitate a sound, but this bird knows the difference between high and low notes and gives each at will.

Then the bird thought he would have some fun with the visitor. He flew up on her hand and gave several digs at it with his bill. He looked amazed when she showed no fear, and tossing his head exclaimed,
" **Go away!**"

The visitor did not move, and the bird seemed to think a minute and then said,............... " Gehe weg."

As this command in German had no effect, Mino apparently decided that his stupid guest would surely understand French, so he called out, ..." Allez-vous-en."

After making a great ado about it the little fellow sang with the combined gift of a canary and a mocking bird. When it was too late Mino concluded to be good and to show off all his accomplishments. As soon as he found that Miss Thursby was going to shut him up in his cage and hurry away to a musical he was willing to do everything and anything that she asked, and as she and the visitor disappeared he called out,   *" Mamma., Mamma.,   please let me go out, I'll come right back."*

But He Uses Naughty Words.

" Once when Mino was visiting me," said to a friend of Miss Thursby's, to whom the bird is sometimes lent, " a naval officer called, and hearing him singing a Chinese funeral song, asked. Does he ever use bad words?"

" No, I answered emphaticaJly. ' It is remarkable, but Mino only uses the choicest language.' Then he called the bird to him. Mino came and bowed before him, and the officer spoke to him in a foreign language. Mino straightened up and answered glibly something I'd often heard him say. The officer laughed and said to me 'He says to me.' "Oh, you English devil."

The bird speaks Chinese supposedly, by the hour. Not speaking the language much ourselves, we aren't sure just what it is. I have never known him to learn anything new since he came to this country, unless taught by a very pretty, light blonde woman. I take it from that that he is a man of the world.

Mino as a Humorist

Mino always anticipates a joke and laughs out before the point is reached. A well-known I archeologist conceived a perfect hatred for the bird, because as soon as he came in for a visit, Mino would begin to sneer at him in the most disagreeable way. He said that the bird was a fiend and that he would not call at the house twhile the bird was an inmate. He did not.!

The bird can cough like a person in the last stages of consumption, and he found out that it annoyed very much a lady who visited me frequently. After that, the minute she came into the house he would laugh like a little fiend and then begin to cough and gasp in a manner that was positively sickening. On the other hand, he knows how to be very considerate. He never awakens anyone in the morning, but keeps perfectly still until you get up. He has a temper of his own, too, for I remember once it was a cold morning, and he wouldn't take his bath. My husband pushed him into the tub, and the bird refused to speak to him for over a month, and he is devoted to him, too !!

Note: Occurred in the USA

Hill Mynah in Aviary
This was an outdoor flight where shrubs were grown. The short perch is being used, but a longer perch is also provided.

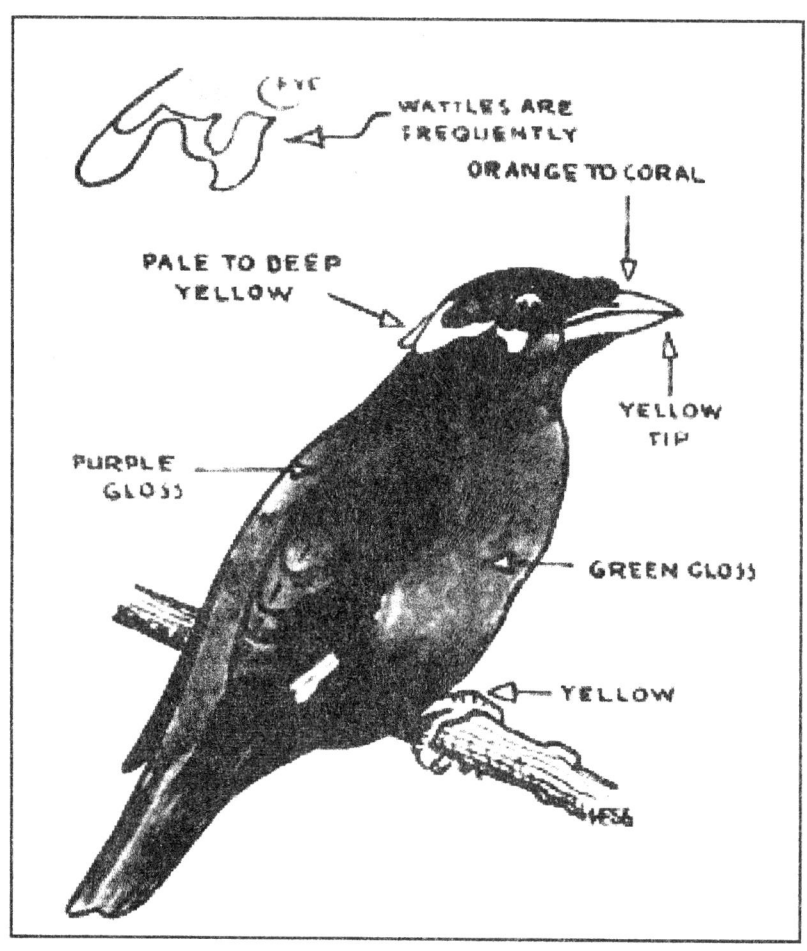

Features of the Indian Grackle (Mynah)
Gracula religiosa

4
FEEDING

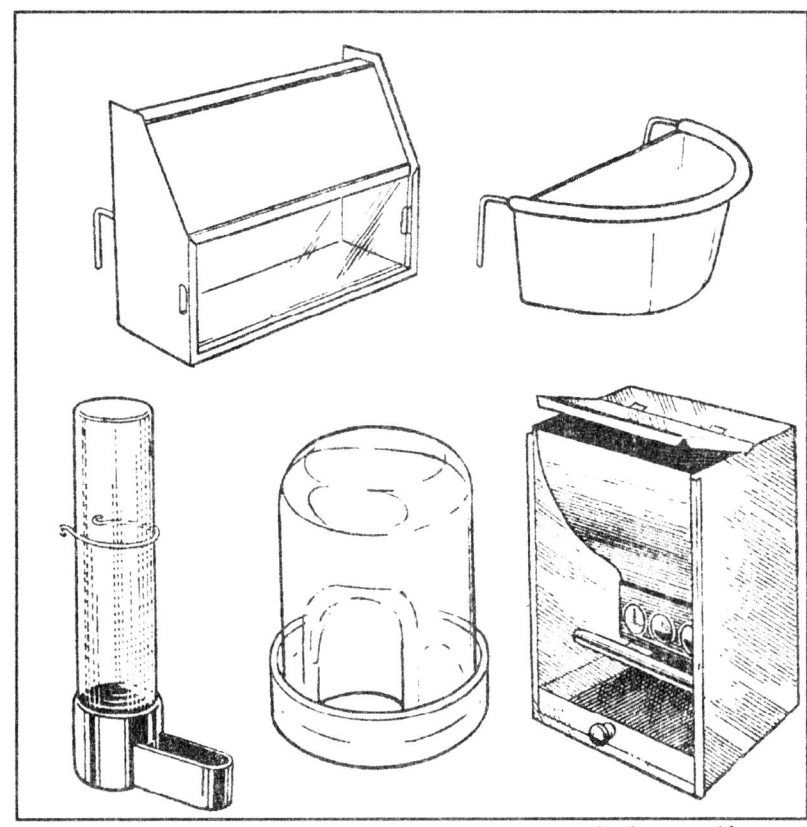

Some feeders and drinkers: *Top left*—A seed hopper that hangs outside the cage. The glass front enables supply to be readily checked. *Right*—A plastic container used as either a feeder or drinker for a cage. *Bottom left*—A simple drinker. Part of the bottom fitting goes through the cage wires. *Centre*—A round water fountain which ensures an uncontaminated and constant supply in a flight. *Right*—Three or four birds can feed at once at this metal seed hopper. The lid is for refilling the container.

Various Types of Food and Water Dishes
Heavy pots would also be used for fruit and mixtures given daily.

PRINCIPLES OF FEEDING

Mynahs belong to the *Soft Billed Group* of birds which means that, unlike the popular finches which eat mainly seed, they take as their normal diet a wide variety of foodstuff. This should come as near as possible to their natural food, although, it will be appreciated, that many of the items taken in the wild may not be obtainable, so substitutes will have to be found.

EXAMPLES
taken from field studies by various writers.

Pagoda   Grass hoppers and other insects. Seeds and buds of pines. Fruits, berries and flower buds.

Ceylon Mynah   Fruits of Ceylon such as wild cinnamon and nutmeg which it eats whole.

From these notes it will be seen that the birds eat what is available in the country concerned. They are very adaptable and spread themselves around to obtain what they can. The common foods are as follows:

1. Grain.
2. Fruit.
3. Flowers and plants.
4. Insects.

In the wild the birds will also pick up minerals and green food, thus making a balanced diet. The problem with aviary or cage birds is how to supply the extra requirements which keep the birds healthy. Obviously, they should be given a selection of different requirements so they can select what they desire.

Examples of Fruit

Banana, Apple, and small Orange. The birds will decide which they like and, remember, any fruit over one day old should be removed. It should be cut up as explained in the text.

MYNAH BIRDS

FRUIT: About 14g of fruit should be fed. This could include: Apple (sliced), bananas, dates, grapes, oranges, pears, plums and any exotic fruit that may be available. Also berries such as elderberries, blackberries and raspberries may be taken. Tinned fruit cocktail or other fruits may also be given.

GRAINS & SEEDS: 5g. Barley, maize, wheat, and bird seeds such as millet or canary.

INSECTS: 5g. Crickets, grass hoppers, meal worms and earth worms. These can be purchased from a specialist supplier (usually advertised in *Cage & Aviary Birds*) and meal worms can be kept for some time in box with bran and slices of raw potatoes for food.

EGG FOOD: 5g. Hard boiled hen eggs, sliced, on which is sprinkled multi-vitamin powder.

OTHER ITEMS: 5g. Boiled Rice, Grated carrots, minced meat, dog or puppy mixture mixed with a little milk. Not all agree with feeding meat, but it does contain essential protein. It is a substitute for the meal worms or other live insects. Cottage cheese (low fat) is also used as a basis for mixing the fruit, along with oat meal, to give a palatable semi-moist mixture. Bread soaked in milk is also fed.

DIET FOR BIRDS KEPT INDOORS

Birds kept in a cage for teaching to talk may cause problems with the scattering of food and the loose bowels if fed food which tends to act as a laxative. More rice, dog mixture, and hard boiled eggs may be a better diet. This alternative diet is often used for the Hill Mynahs, which appear to require more food any way.

NOTE: 1gram = 0.0353   1 oz. = 28.35g

Twice Daily

Birds should be fed twice daily at regular times, preferably first thing in the morning then late afternoon and evening. Soft food quickly turns sour or mouldy so should not be left around to attract flies and possible disease. Accordingly, any food left should be removed after a reasonable period.

If birds are leaving a large amount of food, then too much is being given and the ration should be reduced. On the other hand, very active birds, flying around in an aviary, will require more than cage birds. Also when birds are breeding or are moulting more may be required as well as extra protein or vitamins.

EXCLUSIONS FROM THE DIET

Some bird-keepers cannot stand the idea of keeping meal worms and other live creatures and even the notion is quite off-putting. However, there are variations which may be permitted, provided the birds do not suffer. As noted above, dog meal, oat meal, cottage cheese and minced meat are all good foods. They can be mixed together in a friable mixture and fed in a heavy dish such as that used for small dogs.

The fruit, sliced where appropriate, can be fed in a separate dish. If raisins are given these can be soaked overnight to make them more palatable.

GREEN FOOD

The birds kept in captivity always love to see greenstuff and when nothing special is available they can be supplied with grass, preferably nice and short and quite green. After being deprived of green food they get quite excited and pull at it greedily from the greens' rack, until it finishes up on the floor of the cage or aviary.

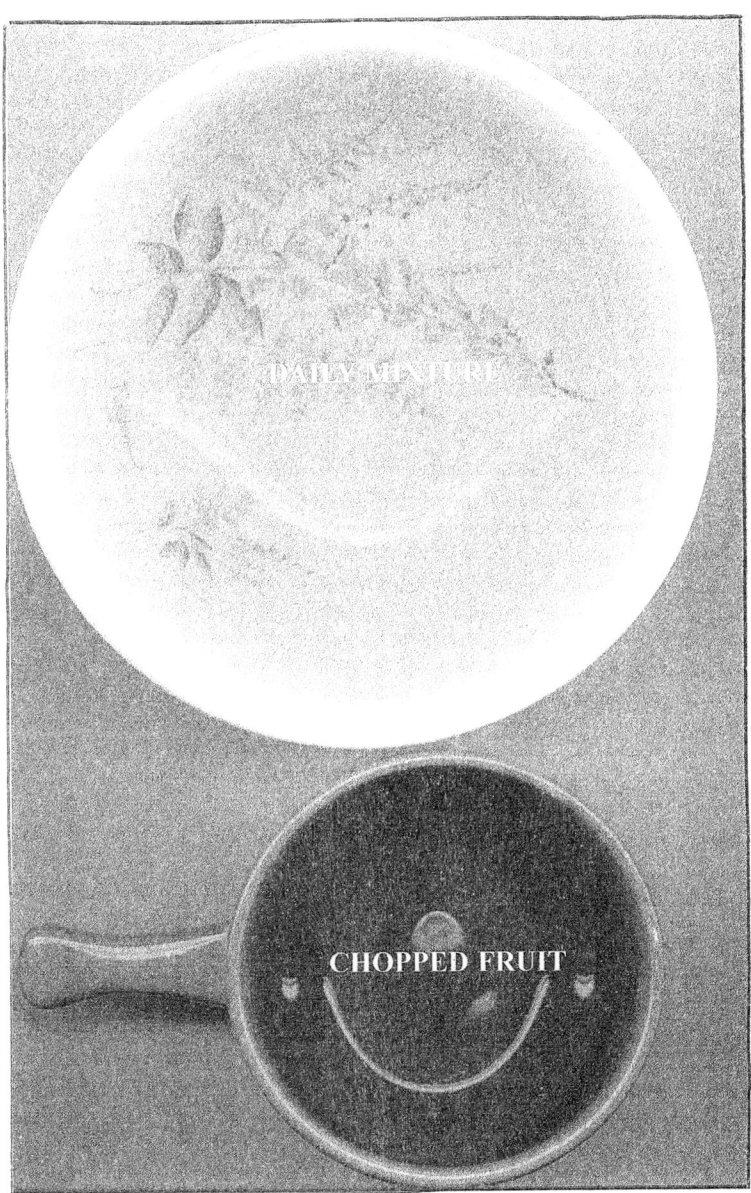

Food Dishes: Heavy type of dishes which cannot be overturned.

There are many tempting titbits which can be found once the Spring arrives. Chickweed, a great favourite, starts to grow when there is sunshine and no frost. After that there is usually an abundance which can be pulled and given to the birds. Some of the typical plants are as shown.

Mynahs will eat most things from soaked bread, wheat, yeast and so on, including any proprietary insectile food available from a pet shop. Such foods as hard boiled eggs, cut into section, grated carrots, cooked peas, raw tomatoes, bananas, skinned and sliced, pears cut into small pieces, and so on, are all acceptable.

Feeding the Gaper

The young Mynah will eat a similar food to the adult, but to make life simple feed Puppy meal mixed with milk and give the finely chpped fruit in a separate dish. Also add cod liver oil and soluble vitamins, but not too much, which is wasteful.

Possible Variations

A household blender or mincer may be used to make preparation of the food easier. This can be used to make a batter using eggs, soya flour, cheese cut in slices, sugar, self raising flour, honey, milk, yeast and so on, so that it becomes a nutritious mixture. Once baked it can be crumbled and kept in an air tight tin so that it keeps for a long period. No quantities are specified because much depends on how many birds are kept, but it should be quite rich in protein from the cheese, soya and other ingredients.

The fruit could be put through the blender and the mixed with any meal used as the basic food and this avoids having to remove the fruit which has become stale, and you can be sure the birds are getting the necessary fruit juice. Mashed potatoes could also be added to make more bulk and butter or minced peanuts could be added.

Fruit and insectile mixture would also be given daily as well as any mixture.

Try to avoid food which causes very loose bowels.

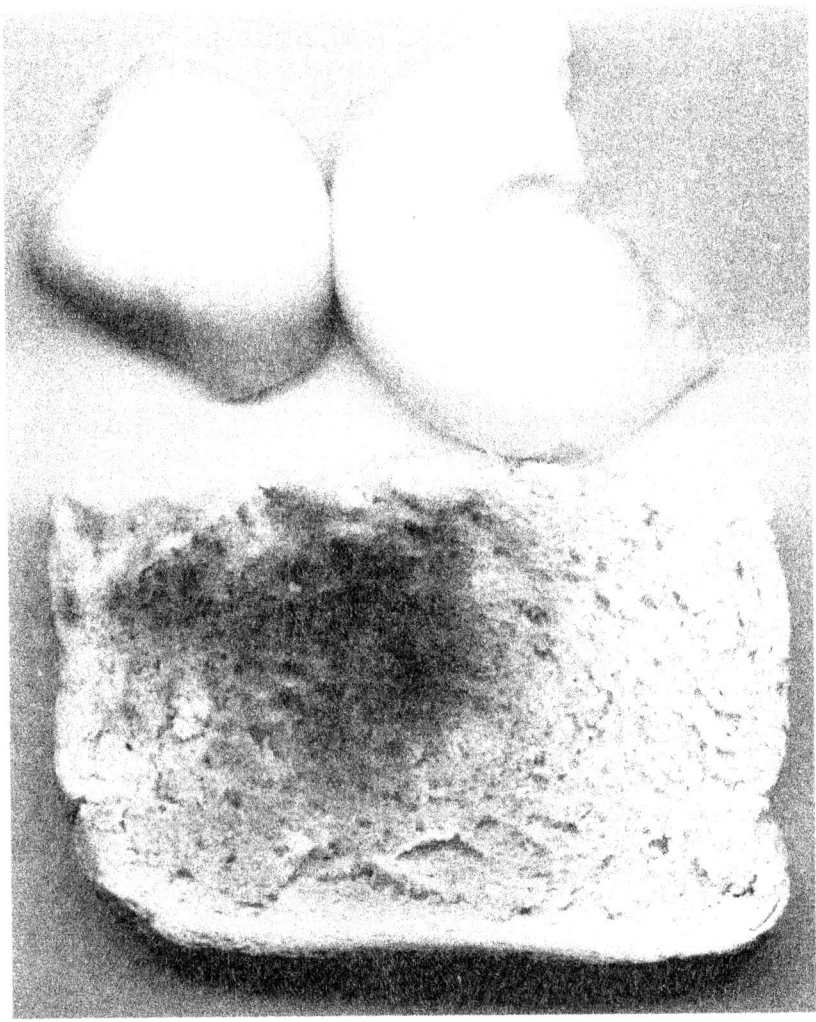

Sections of Hard Boiled Egg. These would be cut into small pieces.
Toasted Bread (leavings from breakfast). Can be soaked in milk and then squeezed out to almost dry.

These are two basic foods, but remember that an active bird like the **Mynah must have sufficient protein so other foods must be given.**

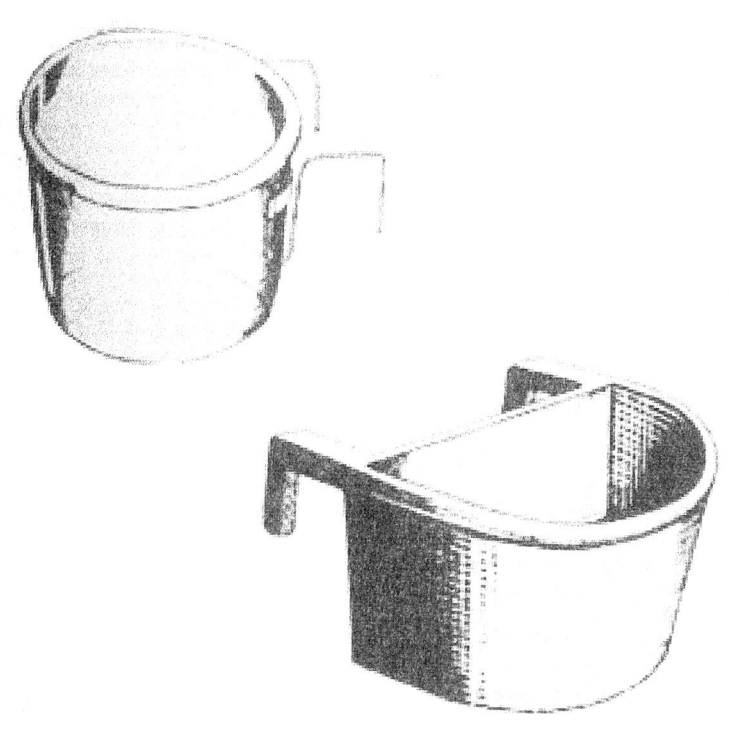

Utensils for Food or Water
An alternative is the tubular drinker (shown earler) which fits outside the cage with the drinking part inside the cage.

CONCLUSION

Feeding Mynahs can be very much a hit and miss affair because they will eat most foods, but these may not always be the correct mixture for a balanced diet. Around 20 per cent protein would be the appropriate rate.

EXAMPLES

Low Protein Foods
Lettuce 1%
Apple 0.50 %
Carrot 1.20 %

High Protein
Egg Yolk 15 %
Egg White 10%
Limseed 23%
Peas or Beans 25%
Wheat 15%
Soybeans 35%
Bread 12% (depends on type)

A *rough idea* of the calculation is as follows:

Egg Yolk	15%
Bread & Milk	12%
Soya	25%
Lettuce & Fruit	2%
	54%

Divided by 4 this gives about 14 per cent. However, this calculation does not take into account the *amount* of food so to be more accurate a weighted average would be used. Thus if bread and milk represented 80 per cent of the food the remaining 20 per cent would have to be very high protein food, sufficient to make up about 6 per cent in the total diet; ie, 20% x 35% protein = 700 divided by 100 = 7 per cent. protein.

CORRECT AMOUNT OF FOOD VITAL

If birds are to be kept in a healthy condition they must be fed and watered in the correct amounts or their health will suffer. This aspect has been summarized by a breeder* as follows:

> I think the main thing to guard against is fatty degeneration, from too rich food and too little excercise! My advice, for what it is worth, would be to feed what they will eat, but not too much fat. Let the bird have plenty of flying space if caged, in the shape of a flight in which it can use its wings.
>
> Good thick perches, which it can grip without straining, should be fixed on with springs at either end. These Grackles, however, seem to stand a great deal of abusive treatment, but, if over fed and " under " caged, sometimes become asthmatic or cattarrhic, their eyes and nostrils discharging a whitish mucous, and their breathing becoming laboured and wheezy.

Fortunately, Mynah birds are disease resistant and hardy so they should present little trouble provided they are housed properly and fed well. If they develop colds or catarrh then they should be fed a proprietory tonic and honey and lemon juice can be added to the drinking water. Only a few drops should be given each day.

---

* H A Fooks, ibid.

5
CASE STUDIES

Common Mynah

THE PAGODA MYNAH

I have always been rather surprised that the Pagoda Mynah has not found more favour with lovers of Foreign Softbills than it has done. Apart from the fact that it has no song worthy of the name, this is an ideal subject for both cage and aviary, and it is a most suitable bird for the beginner.

The Pagoda is about the size of one of our native Starlings. In a quiet way it is very pleasingly coloured, with the breast and under part pinkish fawn, and the back and wings light grey and black. It has a crest of long thin feathers, black in colour, which it can raise or depress at will. Around the beak is a distinctive patch of bright blue skin. Both sexes are coloured alike, which makes them rather difficult to sex.

Its requirements are simple. A good softbill mixture as its staple food is necessary, together with fruit. Ripe sweet apple, pear, and grapes cut up, are all acceptable, and such native fruits as alderberry, raspberry, and blackberry, when in season, may be given. Some live food is essential; cleaned gentles or mealworms, together with any of the numerous insects found in and around the average garden. Although Pagodas certainly enjoy garden insects, they do very well on an allowance of gentles each day.

When acclimatised they are hardy birds, and may safely be kept in a garden aviary all the year round without artificial heat, if a comfortable shelter is provided for them to roost in.

My present Pagoda, has lived without heat for the last three years, and, even during severe frost, has never appeared in the least uncomfortable. Like all Pagodas, she is a great bather, and I have seen her plunge into the water on a cold winter's day immediately I have broken the ice and changed the water.

In addition to their hardiness these birds are usually very long lived. My previous Pagoda, by no means a youngster when I obtained him, lived very happily for about twelve years before finally dying in his sleep from old age.

---

*By R. W. Trippett, (*Scunthorpe*). *Foreign Birds Magazine*, 1955.

I bought him about twenty years ago, out of compassion. I happened to walk into a bird dealer's shop, and saw him in a tiny cage looking absolutely wretched. I did not particularly want him, but I could not leave him, so I bought him for ten shillings. It was only when I arrived home with my purchase that I discovered that he was blind in one eye.

He proved to be a most entertaining and knowledgeable bird. For some reason which I cannot remember he was christened " Boy." He knew and acknowledged his name, and whenever I went to his cage and said " Hello Boy," he would respond by raising his crest, dropping his wings, and spreading his tail, and he would emit a few squawks of greeting. I have, on occasions, put the light on in the bird room in the early hours of the morning, and he never failed to respond to my greeting. Strangely enough he preferred his cage to life in an aviary. During the summer months I turned him out into my aviary, bringing him back into the house in the late autumn. His obvious satisfaction at being " home" again was such that after that I never removed him from his cage. Needless to say he was very tame, and quite often he was allowed the freedom of the living room.

My present Pagoda is a hen, and she has been with me for seven years. Two years ago in the spring, I put a cock Black-Headed Sibia into the same aviary, and after a few weeks the Pagoda and the Sibia became very friendly. They would sit side by side, preening each other, and they accompanied one another about the aviary.

I had a number of Budgerigar nest boxes in the shelter shed, and one day, on inspecting them, I noticed that one contained quite a lot of hay, tufts of moss, and feathers, making a rough nest. Naturally I was rather intrigued, as I had quite a lot of other birds, both Hardbills and Softbills, in the aviary, and I wondered which of them was responsible. Although I kept a careful watch I could never find out which of them were nesting.

After a few days the nest was completed, and an egg (about the size of a Thrush's egg), pale blue in colour, appeared. After two more eggs had been laid I surprised the Pagoda on the nest, and I realised that she was responsible. Altogether five eggs were laid, and I had hopes (not very great I must admit) of producing a rare hybrid.

She sat very closely, but, as I had feared, the eggs proved to be infertile, and eventually she left them. I removed them, and she promptly tried again, laying five further eggs in the same box, but these were also infertile. In each case she sat very closely for over three weeks before deciding that it was a hopeless task.

Last year she again built a nest in the same box, and laid five eggs again, which she brooded for three weeks, but they, too, were infertile. I never saw the Sibia help to build the nest, but he was most attentive to the Pagoda when-ever she came off.

I am on the lookout for a genuine cock Pagoda, as I am quite sure that, if the hen was given a proper mate, she would go to nest, and hatch and rear her young.

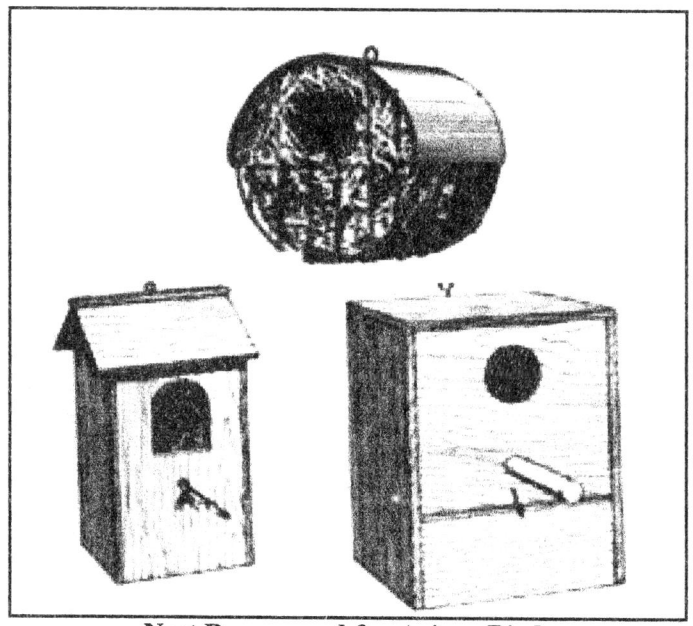

Nest Boxes used for Aviary Birds
Mynahs and Starlings will pair up and use any suitable nest box.

FORMOSA MYNAH BREED

I purchased as a pair in 1957, two Formosa Mynahs, one with a crest and one without. After about two months they both moulted, and both had crests. Had I now got a true pair ? They had one nest of eggs in that year, but they were infertile.

In 1958 they started nesting in my Lovebird aviary, with other Softbills. They killed two Starlings, and destroyed a Cockatiels' nest; I think they ate the eggs. The Mynahs then built a nest of hay, grass, and feathers, and the hen laid four blue eggs; slightly darker blue than those of an English Starling, about the same size, but blunter. They sat for fourteen days, and the young hatched. I gave them fruit, softbill food, and mealworms. The young grew till they were half-feathered, when they died, for some unknown reason.

They had another nest, but the young died when about two or three days old. In 1959, they had three nests, each of four eggs. Some young hatched, but none lived for more than three or four days, perhaps owing to shortage of live food. During both 1958 and 1959, these birds had to fight for live food, against about ten Starlings and other Mynahs.

Late in 1959 I placed the Mynahs (above mentioned) in an open aviary, built of corrugated iron sheets. This had a shelter, with open front. The aviary contained also three Pheasants and a pair of Ringnecks. The Mynahs bossed the lot! The food supply was a softbill mixture, but no mealworms.

This year (1960) they built a nest in an old Lovebird nest box, fourteen inches by six inches by six inches, with a hole in the front, two inches by one-and-a-half inches. This was hung just under the shelter. Later I saw four eggs, but they disappeared after a time (eaten ?), and I gave up hope.

On July 21st, I saw the old birds come out of the box, and looking in it, I found four eggs. On August 4th, I heard young birds calling, so I supplied some maggots.

The old birds came and started feeding. My young son was on holiday, so I got him to fill a deep tin with maggots while I was away. This feeding went on for five to seven days, when I saw the Pheasants rob a Mynah of maggots on the floor, so out came the Pheasants, and the maggot consumption dropped fifty per cent !

On August 25th, I had a look in the nest box. There were two big young birds in it, and one small one. One jumped out, but I put it back. On the 26th one young bird was on the perch, and another on the 27th. The last (the small one) came on out the 29th, but died on September 2nd.

The other two grew fast, and flew well. About September 10 th one appeared to have been killed. So I caught up the last young one, and put it in a cage. This is now doing well, and feeding on ninety per cent softbill food, with a little apple and an odd maggot.

The young bird is like the parents, but dull brownish black, and with no crest. It has the white wing bars, which are seen when flying, and its legs are reddish. Its beak is the same as the adults.

NOTES ON CASE STUDY (AUTHOR)

A number of lessons can be taught from this case study. The main points are as follows:

1. Feeding

It would appear that the feeding was not properly balanced. Possibly too much importance was attached to feeding paggots.

2. Mixed Collection

If attempting to breed there is much to be said for keeping one species in the aviary. Mixing other birds is most likely a source of much trouble.

---

* Based on the experiences of G W Bratley, in *Foreign Birds*, February, 1961.
Mr Bratley was awarded a Certificate for this achievement by The Foreign Bird League.